Video Analysis and Repackaging
for Distance Education

A. Ranjith Ram • Subhasis Chaudhuri

Video Analysis and Repackaging for Distance Education

 Springer

A. Ranjith Ram
Government College of Engineering
Kannur, Kerala, India

Subhasis Chaudhuri
Indian Institute of Technology, Bombay
Powai, Maharashtra, India

ISBN 978-1-4899-8577-4   ISBN 978-1-4614-3837-3 (eBook)
DOI 10.1007/978-1-4614-3837-3
Springer New York Heidelberg Dordrecht London

To

Dearest Gita, Kamal and Kirtana

ARR.

Raj, Shankar and all my teachers

SC.

Preface

The tertiary sector of our economy requires that a significant portion of the population be well educated. However, a vast majority of the population in developing and under-developed countries do not have a proper access to higher education. Higher education is expensive and, quite often, there are not enough qualified teachers in the locality. Education is a human-intensive training program and we require a very large number of well qualified teachers. Further, the progress in technology is very rapid today and our work-force is amenable to quick obsolescence unless we provide a continuous education program to the working professionals. This requires that we allot a significant amount of resources to our national educational program, which most of the countries are unable to do. As the demand for the higher education grows, so is the restlessness among the young generation to have access to quality education.

Fortunately, there has been a massive expansion in distance delivery and e-learning systems over the last decade. Many universities now offer their courses live to participants across the globe, and even offer complete or limited interactions with the instructor. Several e-learning portals offer video-on-demand services for specific courses. Many universities make their course materials available to the outside world for the benefit of all. One such laudable effort is the NPTEL program (http://www.nptel.iitm.ac.in) of the Government of India, that promises to make the entire curriculum of each of the undergraduate engineering programs freely available to the entire mass. Whether free or not, distance education does serve the purpose of enhancing the outreach of quality education to all.

Two important aspects of distance education are the quality of contents and the mode of delivery. It is imperative that the quality of contents be good, failing which distance education serves no purpose at all. The delivery of the content is also equally important as there should be no loss in the pedagogic value. A good quality video requires a significant amount of bandwidth for data transmission. In developing and other countries, the available bandwidth is still quite limited. This requires a significant reduction in data during the content delivery. Also, considering the fact that the penetration of mobile phones even in rural areas is very high, we should also explore the possibility of delivering the contents through the mobile display unit. This puts a further constraint on the bandwidth during the transmission of the lecture video. Another added constraint is the limited size (resolution) of the display unit. One should be able to deliver the content on smaller display units without sacrificing the legibility of the contents.

The purpose of the monograph is to explain how one can enhance the outreach of distance education by appropriately repackaging the instructional video so that the amount of data can be drastically reduced without sacrificing the pedagogy. We have built a complete system that takes hours of classroom lecture video, processes it to generate a compressed representation of the data, called the instructional media package, and can play it on diverse multimedia platforms, including mobile phones. We believe that such a monograph is very timely as development of efficient distance education platforms is the need of the hour. We discuss various components of the system in full details with a view that practitioners of this area will be able to benefit from this monograph.

This monograph is an outgrowth of the Ph. D. dissertation work of one of the authors under the guidance of the other at the Indian Institute of Technology Bombay. The book is a revised and extended version of the thesis. Several of the ideas presented in the monograph are quite novel and have been filed for patents. A fully functional android platform based media player, called *Lec-to-Mobile*, has been released to the public in 2011 and is freely downloadable form the android market site. We recommend that the readers use this player on a mobile display unit for a better understanding of the discussed topics.

This monograph may be of value to video analysis researchers and practitioners who are interested in developing technologies for video analyses in general, and educational video in particular. The book is mostly self contained and is intended for a wide audience. The mathematical complexity of the book remains at a level well within the grasp of undergraduate students. A basic familiarity with the area of image processing should suffice. Hence the students may find this book useful as a reference. We have provided a large number of figures to help understand the topic well.

We welcome comments and suggestions from the readers.

Mumbai,   *A. Ranjith Ram*
January 2012   *Subhasis Chaudhuri*

Acknowledgments

The effort to write a book suffers from a periodic nonchalance and is impossible without the active support and encouragement of a large number of people.

The authors are thankful to Prof. Sharat Chandran and Prof. Shabbir N. Merchant at IIT Bombay for their insightful comments and ideas at various stages of research over the last four years. Thanks are also due to Prof. Jayanta Mukhopadhyay of IIT Kharagpur, Prof. John R. Kender of Columbia University, and Prof. Venkat Rangan of Amrita Vishwa Vidyapeetham for their valuable comments. The help given by Trupti N. Thete in implementing the audio watermarking and Anup Shetty in implementing the optimal video shot detection is greatly acknowledged. The authors are also thankful to Prof. Kannan Moudgalya and the Centre for Distance Engineering Education Programme (CDEEP) at IIT Bombay for providing various instructional videos. A few undergraduate students took up the task of developing the Lec-to-Mobile player. The authors thank Himani Arora, Deval Agrahari, Manas Chaudhari and Siddarth Sarangdhar for the same. The authors are also indebted to IIT Bombay for its effort in getting the developed technologies patented.

The first author wishes to acknowledge the support received from the Quality Improvement Program (QIP) to carry out research at IIT Bombay. The second author acknowledges the funding support from the Department of Science and Technology (DST) in the form of a JC Bose National Fellowship. The administrative support from the publisher is also gratefully acknowledged.

The authors are thankful to their family members for their love, support and constant encouragement.

Mumbai,   *A. Ranjith Ram*
January 2012   *Subhasis Chaudhuri*

Contents

Chapter 1
Media for Distance Education

1.1 Introduction

Education can be thought of as the process of adding value to life. Apart from its fundamental goal of gathering knowledge, it has manifold long term objectives. One may say that education focuses at (i) value creation, (ii) economic viability, (iii) broadening of perspective, and (iv) enhanced social outreach. It should help an individual mold his/her personality, earn a living, which in turn accounts for his/her social health and lifestyle. It is hard to define the efficiency of educational systems from the value creation point of view since objective measures of value are not possible. In order to devise some means to quantify the efficiency of education, it would have to be treated more like a commodity rather than an abstract concept of knowledge. In this aspect, educational systems may be modeled as commodity markets whose performance measure is given by the throughput of the system. We define the throughput to be the average number of students graduating per year per instructor for a given curriculum.

A quick tour on the evolution of education systems would be helpful to note how their transformation has affected the throughput over years. About 3000 years ago, the education system was based on *gurukula*, which means the extended family of the *guru* (teacher) wherein the students reside until graduation, help performing the household chores and learn from the guru. It did exist in India, China, Japan and Greece in which a disciple was resident with his/her *guru* or teacher. That is, the school was primarily residential in nature in which all students entered at a lower age, lived with their teacher and left the school when the education was deemed to be complete. Both the admission and graduation processes were asynchronous in nature. The disciple learnt everything from his/her *guru*, including the day-to-day needs like cooking, washing cloths, and handling weapons. As per the education model already mentioned, this system could be thought of having a single *vendor* (one *guru*) and multiple *customers* (M disciples). The period a particular student had to spend in the *gurukula* to graduate depended on the competence of the student as well as the process of instruction followed by the guru. The throughput of such a system could be assumed to be about 0.5 in the absence of any appropriate documentation. Hence on an average, one student graduated every two years from a guru and this is naturally quite low. However, given the nature of the economy in those days, this was probably good enough as the tertiary sector of economy was yet to arrive.

Around 1000-2000 years ago, there evolved *universities* as port-folio managers in education. The examples of such a system are Taxila University and Nalanda University, with which N *gurus* and M disciples were associated. Again the throughput was found to be almost similar, but with the possibility of providing a better domain knowledge since a student is in touch with multiple teachers under the university during the graduation process. The scenario changed very much from 14th century onwards due to the advent of well structured university systems in Europe. The same model of education was driven by renaissance and industrial revolution. Technology evolved as a subject in these universities during this era. Also gradual streamlining of the university system took place and the throughput could be increased beyond 2.0^1. The teacher to student ratio slowly increased to about 1:10.

After 1990, there was another noticeable change - the model remained the same, but with a more information dissemination capacity. The key facilitator was the Internet and the associated digital technology, by which the outreach of education is drastically increased. There evolved a *distance education system* and the teacher to student ratio could be further increased up to 1:30. One could attain an enhanced system throughput of greater than even 5.0 in some of the universities now. For example, at Indian Institute of Technology Bombay, the throughput is currently 3.0 while in Europe and in USA, some of the non-research intensive universities have attained a throughput as high as 6.0. Lately many of the universities are also experiencing financial difficulties due to reduced or limited state support and the universities need to enhance the throughput for better financial solvency. But how does one increase the throughput? The answer possibly lies in how efficiently we can enhance the outreach of distance education.

1.2 Distance education and e-learning

We have explained the need for distance education in the previous section by substantiating the increase in throughput while extending education beyond classrooms. With the proliferation of various new technologies in the field of distance education, the concept of *teaching* and *learning* has undergone revolutionary changes in the last one decade. In this aspect, real classrooms are being augmented or even replaced by an e-learning environment. With this, a student can use a computer and communication networks for off-classroom learning. Recently many distance education providers also came into picture, who either perform a live tele or webcast of the class room being taught or supply the DVDs of the

---

[1] Data in support of such a number is not available. The authors used their subjective judgment to arrive at the numbers.

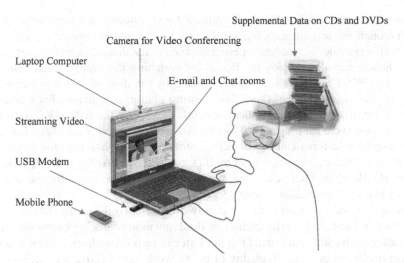

Fig. 1.1 A diagram showing the different technology interfaces used by a student in an e-learning environment.

stored lecture video or maintain video on demand for different courses. Examples of such educational systems are the Distance Education Technology and Services (DETS, website: http://www.dets.umd.edu) by University of Maryland, Division of Continuing Education (DCE, website: http://www.dce.ufl.edu) by University of Florida, School of Continuing and Professional Studies (SCPS, website: http://www.scps.nyu.edu) by New York University, the Open University in U.K. (website: http://www.open.ac.uk), Centre for Distance Education (CDE, website: http://distance.uaf.edu) by University of Alaska, and Centre for Distance Engineering Education Programme (CDEEP, website: http://www.cdeep.iitb.ac.in) by IIT Bombay.

Now-a-days the Internet also provides an effective means for web based education and infotainment. The student can also attend examinations *online*. Examples of such educational systems are Stanford Engineering Everywhere (SEE, website: http://see.stanford.edu) by Stanford University, Distance Education Network (DEN, website: http://den.usc.edu) by University of Southern California, Northeastern University Online (website: http://www.northeastern.edu/online), University Extension of the University of Texas at Austin (website: http://www.utexas.edu/ce/uex/ online), Boston University Online (website: http://www.bu.edu/online), and Arizona State University Online (website: http://asuonline.asu.edu). The spread of such educational systems have opened up new dimensions on teaching and learning processes such that a student can use the learning aids *anywhere, anytime*.

A typical e-learning environment is shown in Fig. 1.1. The student is in interface with the knowledge world through a laptop computer equipped with Wi-Fi LAN and through cellular telephony. Here the pedagogy is effected entirely by electronic means, and hence is the name *e-learning*. User can browse the Internet, participate in a video-conference or play video on demand on the Wi-Fi networked laptop.

IP telephony can also be effectively utilized by the student for knowledge sharing. Through the technologies like GPRS, Wi-Fi, 3G and beyond, the cell phone would also provide the student an efficient means for distant access. Apart from these, he/she can also employ the laptop for watching the supplemental material on the CD-ROM or DVDs which are provided by the distance education centers. One can also utilize many of the online learning platforms available. For example, in India, for distance learning students can effectively use the A-VIEW$^{®}$ (Amrita virtual interactive e-learning world) tool by Amrita Vishwa Vidyapeetham. Further, supplemental e-learning can be effected through NPTEL (National Programme on Technology Enhanced Learning) run by IITs and IISc, and AAQ (Ask A Question) run by IIT Bombay. These electronic means are found to be very efficient and attractive in the dissemination of knowledge.

The quality of education always lies in two factors - (i) the depth of knowledge of the teacher and (ii) the effectiveness of dissemination of his/her knowledge. The first factor is absolute and crucial but the latter in turn depends on several aspects like (a) quality of oration, (b) quality of board work, (c) effective interaction with the students in the classroom, and (d) the body language. As we have already explained, the advancements in communication technology have largely contributed to the both ends of knowledge sharing - *teaching* and *learning* during the last few years. Consequently the outreach of education has been extended from the classroom to distant places so that the accessibility of knowledge centers has improved. For such augmented learning methods, video is found to be very powerful due to its inherent ability to carry and transmit rich information through its constituent media.

Digital video is a versatile media with applications varying from entertainment to education. It plays a major role in constituting a knowledge database, irrespective of the domain of usage. The information contained by the digital video is conveyed to the end user by the visual data present in the video frames along with the associated audio stream. This accounts for a high bandwidth requirement for video transmission. The visual patterns present in image frames usually contain a large amount of redundant information which bring about another limitation of a large memory requirement. Therefore the price paid for the efficiency in conveying information by video includes both memory and bandwidth. Hence the content analysis of digital video by which it can be represented in a compact form is inevitable for its easy access and fast browsing. In the present scenario of the increased use of communication networks and the Internet, the research and development of new multimedia technologies which aim at structuring, indexing, summarizing, meta-data creation and repackaging, would really contribute to the field of infotainment. Our effort in this monograph is to suggest how these tasks can be achieved. To start with, one should study the organization of a digital video and the related nomenclature first.

1.3 Organization of digital video

The term video commonly refers to a series of temporally correlated visual and audio data streams presented in a synchronized fashion. In digital video, since the constituent media appear in a digital format, the frame sequence comprises of digital images and the audio stream comprises of sampled audio data. On storage, these are combined and saved in a single file to form a digital video data. Several video file formats are popular now-a-days, like AVI, WMV, ASF, MOV, MPG, RM, FLV, SWF and 3GP. The AVI (Audio/Video Interleave) format uses less compression than other formats and is very popular. WMV (Windows Media Video) format was developed by Microsoft, which was originally designed for Internet streaming applications, but now can cater to more specialized contents. ASF (Advanced Streaming Format) is a subset of the WMV format. It is intended for streaming and is used to support playback from digital media and HTTP servers, and to support storage. The MOV (Quicktime Movie) was developed by Apple and is also very common now-a-days. MPG (Moving Picture Experts Group) format uses MPEG standards for audio and video compression and hence caters to the needs of majority of users. RM (Real Media) format was created my RealNetworks, is typically used for streaming media files over the Internet. FLV (Flash Video) format has become increasingly popular due to the cross-platform availability of Flash video players. Flash video is compact due to compression and supports both progressive and streaming downloads. SWF (Flash Movie Format) format was developed my Macromedia, which can include text, graphics and animation. They could be played once Flash Plug-in's are installed in web browsers. 3GP (3G Cell Phone) format was designed as a multimedia format for transmitting video files between 3G cell phones and the Internet. It is commonly used to capture video by the cell phone camera and to place it online. Hence we observe that a large variety of options do exist for the universities to offer their distance education program. However, one should not select any format that restricts the use of the media to the masses. The choice should be such that it is platform independent and is efficient in compressing the data.

The visual and audio information are interlinked and stored appropriately in these file formats. The synchronization between these two media is so important that a deferred sound track would result in semantically incorrect perception. The stored video frames are delivered at a fixed frame rate irrespective of its content, to render a feel of continuous motion on the screen. Before going further into the constitution of video, it is better to review some of the related terminologies which will be used in this monograph.

Frame: A single static image taken from a physical scene is called a frame. A number of frames are to be captured per second so as to preserve the continuity of motion in the scene. This rate at which frames are shot by the camera is called the frame rate. A typical value of the frame rate is 25 per second.

Shot: A shot is a temporally contiguous frame sequence with the characteristic of motion continuity which may result from an uninterrupted camera or scene motion. In other words, it is a stream of frames recorded continuously depicting

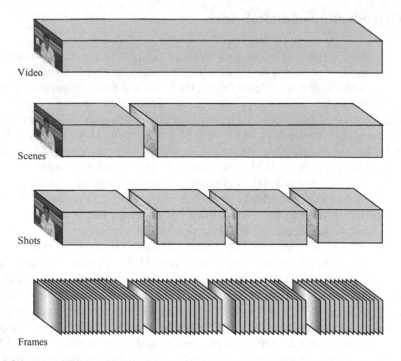

Fig. 1.2 A schematic diagram showing the composition of video data.

a physical world. Several shots can be sequentially connected by editing to create a video.

Scene: A scene is a sequence of shots related through some semantic features. The content of a scene should obey the boundedness properties in space, time and action. All shots sharing such a property are part of the same scene. The organizational structure of a video in terms of scenes, shots and frames is shown in Fig. 1.2.

Motion: It is the movement of pixels from frame to frame. Camera motions like pan, zoom, etc., result in a *global* motion and moving objects in the physical world result in a *local* motion. If motion is less, one can say that there is a high temporal redundancy in the video.

Activity: Activity refers to what is being performed in a video segment. It is treated as a combination of the content in a frame and its motion throughout. Usually a single shot contains one video activity and hence sometimes shot detection is often related to activity detection in video.

Key-frame: A key frame is a characteristic image frame which represents the visual content of a video segment. A long-lasting video sequence can be represented, in short, by a temporally ordered key-frame subsequence. The number of key-frames present in a video is entirely determined by its scene composition.

Fig. 1.3 A typical classroom scene. Note that the scene complexity of the resulting video shots would be very low.

There are a wide variety of digital videos available and the analysis methodology is highly domain dependent as different classes of videos like news video, sports video, music video, educational video, etc., show an organizational variation in terms of their scene composition. A domain specific analysis designed for a particular class of video may fail to perform adequately for another class. So a comparison of an instructional video with other categories is necessary before attempting a content analysis of the former.

1.4 How is a lecture video different?

The lecture video produced at a live classroom differs in several aspects from a commercial video. A commercial video is comprised of several shots from different scenes, with the associated speech signals originating from different speakers. These shots may also have severe camera motion like pan and zoom, and also contain different objects in motion. On the other hand, a lecture video is produced inside a live classroom or at a distance education studio by lightly trained staff who is required to do less camera work and editing. A typical classroom scene is shown in Fig. 1.3. Note that the types of shots arising from such a scene will be limited to three or four including the electronic slide, black/white board, instructor's face or the audience (students). Hence an instructional video lacks much of motion information, but is rich with textual content, written on slides or on black board along with very rich audio contents. The general definition of *shot* as a temporally contiguous segment of video captured by a single camera cannot be used as such in the case of lecture video. There will be just a few fixed cameras which dwell long and continuously on the instructor, the black board, or other media, regardless of the changes in textual content. Hence typically an instructional video contains a talking head, several slide transitions and/or hand-written portions on a black board or a white paper. Here we redefine a shot as a sequence of frames that capture the same electronic slide or handwritten page. We assume that a lecture video consists of several instructional shots, each of which can be characterized by a single key-frame. With this assumption, we find that there are less than a few tens of instructional scenes in a typical one-hour class room lecture. The work outlined in this book deals with the processing of educational video in an attempt to enhancing the outreach of distance education.

1.5 Motivation for the book

The literature presents many summarization methods for content specific videos but there is no work reported on content re-creation aspect of the video from these summaries. In the present scenario of education being de-centralized through distance learning programs, the technology for instructional media content re-creation from instructional summaries can offer promising goals without much degradation in instructional values. The outreach of distance education can be even extended from remote computers to mobile phones by adopting suitable retargeting strategies for the instructional media. The server at the distance education provider can perform content analyses of the original lecture videos and produce the multimedia summary. This summary can serve as a *seed* for the content reproduction at the deployment side of the distance education. The automatic content analysis, the production of multimedia summary and the corresponding media re-creation find a very potential application in distance education and community outreach programs by which the issues related to storage and bandwidth requirements of lecture videos can be effectively solved. These enhanced technologies can efficiently handle the production, storage and use of instructional media since all the methodologies aim at processing the video by preserving its content.

Distance education centers of universities or commercial institutions have a large corpus of instructional videos of the offered courses. These instructional videos are to be provided as learning aids to students for an efficient dissemination of knowledge. There are however several issues while dealing with a vast variety of instructional videos. First one is on the compression aspects of these videos to save memory and bandwidth, as already mentioned. For this, one may target at the redundancies in the video to eliminate them to provide a compact multimedia representation. Here we come up with a novel idea of *repackaging* a lecture video to produce an *instructional media package* (IMP), from which the media can be re-created when needed at the client site. A student is able to learn by watching the re-created media without any apparent loss in pedagogic content. Then secondly, in an attempt to enhance the learn-as-you-move experience, one can think of developing an appropriate multimedia technology by which these media can be adapted for display on miniature mobile devices. Since mobile phones and PDAs have become pervasive among students, there is a high demand for viewing lecture videos on such devices and to get a quick access to educational content. This can be achieved by using a suitable *retargeting* method which aims at retaining the legibility aspect of the written text in the video frames. Retargeting is the process of transforming an existing signal to fit the capabilities of an arbitrary device. It aims at preserving the users' perceptual experience by suitably maintaining the importance of information contained in the signal. Retargeting of an instructional media on mobile and miniature devices adds to the emerging field of technology in the are of *infotainment*.

The third aspect is that there should be some copyright protection for the repackaged media by which the producers could benefit along with the achieved goal of spreading education. Content protection is crucial in educational systems and hence the components of the repackaged media are to be appropriately protected in this respect. The final issue is that there may be separate videos of the same course by different instructors, which necessitates a *preview* of videos. These previews should be given to the students freely, before attending the course or buying the entire course video. In this regard, we propose to produce a *lecture video capsule* (LVC) which contains the instructional highlights of a given lecture. All these facts have collectively contributed to the motivation for this book.

The various chapters in this book are organized based on the different objectives and approaches for the repackaging and content re-creation for instructional video. In this chapter we presented the introductory concepts related to distance education and instructional videos. Next chapter gives an overview of video processing and repackaging. The current state of the art related to this research area is discussed in Chapter 3. The comparison of the technology found in the literature along with their shortcomings are presented in this chapter. Chapter 4 discusses the preprocessing required for the analysis of lecture video. The issues on segmentation of instructional video and activity recognition for the exploration of the organizational structure of instructional video are dealt with in detail in this chapter. The pedagogic repackaging and content re-creation of instructional video which aim at a drastic reduction in memory and bandwidth is presented in Chapter 5. Chapter 6 extends the content re-creation of instructional media to mobile devices by which a legibility retentive retargeting is achieved on miniature screens. Chapter 7 introduces the watermarking of the instructional multimedia documents for copyright protection. In Chapter 8, a method of content and quality based highlight creation in instructional video is presented. Finally, the conclusions drawn from the various methodologies adopted for content re-creation of instructional video along with some directions for future research are presented in Chapter 9.

Chapter 2
Instructional Media Repackaging

2.1 Introduction

Educational videos serve as a vast collection of information resource which can be used as supplemental material by students, in addition to regular class room learning. Hence the research and development in new multimedia technologies which focus on the improved accessibility of the huge volume of the stored educational video data have become quite crucial. In the distance education scenario, the instructional video should be represented with a less amount of data so that it can be transmitted through a lesser bandwidth. The techniques to achieve this goal should effectively deal with the redundancies present in the instructional video, thus yielding a compact multimedia representation of the original video, without much loss in information. Although recent research has contributed many advances to developing software for education and training, creating effective multimedia documents still remains primarily a labor-intensive manual process. To automate the content analysis or structuring and summarizing an instructional video, major research issues include the detection of slide transitions, detection of text regions, recognition of characters and words, tracking of pen or pointers, gesture analysis, speech recognition and the synchronization of video, audio and presentation slides.

We know that the feel of video is through the illusion of continuity created by delivering about 25 image frames per second along with the audio. The organizational structure of digital video in terms of scenes, shots and frames has already been explained in Section 1.3 of the previous chapter. However our raw data are instructional videos, some specific characteristics of which are also given in the Section 1.4 of the same chapter. We infer that typically an instructional video contains a talking instructor and several slide transitions (handwritten/electronic) and/or hand-written portions on a black board or a white paper. Those image frames which are rich with textual contents constitute *content* segments in the video and those frames in which a talking instructor appears constitute *non-content* segments as they do not convey any textual content. One may note that much of the image frames in the content as well as non-content regions in the instructional video are redundant in nature as

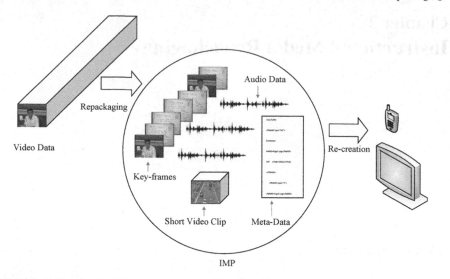

Fig. 2.1 A diagram showing the process of video repackaging and the subsequent content re-creation.

they either represent the same instructor or depict the same textual content. Since the variety in instructional scenes is limited, the key-frames that represent the original video will be very few in number. If we are able to extract these key-frames along with the information about their temporal location in the video, we can devise a mechanism of re-creating the video from these data, along with the complete audio. Hence in an attempt to solve the issues related to the high memory and bandwidth requirements of instructional videos, we come up with a novel way of representing the original video in terms of its key-frames, the associated audio file and the related meta-data.

Media repackaging means representing the media of interest in another form, without much loss in generality. That is, a listener or viewer should perceive the altered media in a semantically organized way and with a reasonably good quality. For the repackaging of instructional video we aim at picking up (i) relevant key-frames, (ii) the complete audio file, and (iii) a video meta-data which controls the playback. We use the terminology *instructional media package* (IMP) to represent them collectively. The process of video repackaging and the subsequent content re-creation is illustrated in Fig 2.1. On the analysis side, this approach aims at the summarization of the instructional videos to produce multimedia documents like images, audio files, text files, etc. On the media re-creation side, one can view a rendering of the original video from these summaries with an appropriate playback mechanism. Both video summarization and media re-creation are found to be very effective in knowledge dissemination since it overcomes the limitations of memory and bandwidth. Occasionally an instructional video includes video-based demonstration or illustration (imported video segments), apart from the instructional scenes already mentioned. Then these video clips need to be retained as such in the IMP for content recreation. In such a case the IMP not only contains the above three categories of media, but also short video clips as shown in Fig 2.1.

The content re-creation process may be extended to miniature devices having small displays which is also discussed in this monograph. In doing so, the legibility aspect of the text-filled regions should be preserved so that the user (student) is able to view the content with the fullest possible resolution. Another processing task that could be incorporated in the repackaging process is the content authentication. That is, appropriate copyright protection mechanisms may be employed before the deployment of instructional media. Also, to help a student in selecting the desired instructional package from a vast variety of available ones, preview videos could be produced and supplied freely on the web. All of these form the various functional units of the suggested system for the repackaging and content re-creation of instructional video.

2.2 Objectives for repackaging

The overall objective of this monograph is to present various methodologies for the technology enhancement for distance education which aim at the automatic content analysis of instructional videos to produce their multimedia summaries at the server side and the content re-creation at the client side. Towards achieving this goal, the following detailed objectives have been addressed:

(a) Analyze instructional videos to produce a repackaged representation namely the IMP with focus on reduction in memory and bandwidth of transmission.
(b) Discuss a method for instructional video content re-creation from IMP.
(c) Suggest a legibility retentive retargeting of the IMP on miniature mobile devices by which the student can watch the media *anywhere, any time*.
(d) Apply digital watermarking schemes for the IMP for copyright protection.
(e) Perform the highlight creation in an instructional video to produce a *lecture video capsule* (LVC) in order to provide a quick preview for students.

The overall system designed for the repackaging and content re-creation of the instructional video consists of two functional parts namely the server and the client. On the server side, the analysis of instructional video is performed to produce the multimedia summary namely the IMP, followed by its digital watermarking and also to create the LVC. The client, usually a student with a computer or with a mobile phone in an e-learning environment receives the above data for content re-creation. The functional overview of the suggested methodology is shown in Fig 2.2. The original video is first split into its constituent media - the audio stream and image frame sequence. These are then processed separately as shown in Fig 2.2. The video content analyzer derives information required for (a) key frame selection for content summarization, (b) meta-data creation for content reproduction, and (c) video clips selection for capsule generation. The meta-data contains the temporal markings required for the display of key-frames during media re-creation. When the

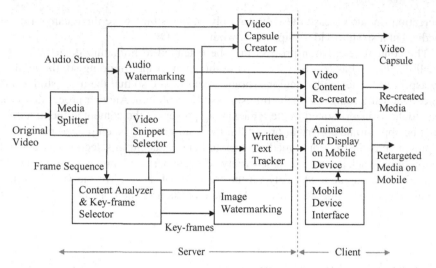

Fig. 2.2 A schematic diagram showing an overview of the complete system for the repackaging and content re-creation of instructional video.

media re-creation is meant for miniature devices, this meta-data contains additional (spatial) information to select the region of interest (ROI) from the content key-frame. The key-frames and the audio are watermarked for preserving ownership rights and are fed to the content re-creator at the client side, along with the meta-data. An estimate of the original video is re-created at the client device, without any apparent loss in instructional values. If the client happens to be a mobile phone, a text tracking meta-data is necessary for a legibility retentive display of the visual content of key-frames. This is because we aim at an ROI based visual delivery on the miniature screen, in which the selection of the spatial window is based on the writing sequence of the instructor in the original video. The capsule creator generates a short preview video, intended for the users of distance education. With the help of preview capsules a user can select a particular instructional package according to his/her requirement. This capsule is actually produced by selecting appropriate video snippets from the original instructional video based on a *highlight* definition strategy. Now we discuss each functional element in detail.

Prior to higher level content analysis, a lecture video is to be temporally segmented and its shots are to be classified. We first discuss the methods for shot detection and recognition separately. The shot detection method is based on the histogram difference and the shot recognition method is an HMM based one. Then we discuss a joint shot detection and recognition method in which the above HMM is configured for detecting the transition in scenes, thereby separating them at the change points so that the individual activities can be continuously recognized with a minimum delay.

As already mentioned in the previous section, in view of solving the issues related to the high memory and bandwidth requirements of instructional videos, we represent the original video in terms of its multimedia document summaries namely the IMP which essentially contains the key-frames, the associated audio file and the related video meta-data. Key-frames can be thought of as a temporally ordered subsequence of the original video sequence that represents the visual content of the video. They are of great importance in video indexing, summarization, content retrieval/re-creation and browsing. In the developed system, key-frames serve as the building blocks for the content re-creation of instructional video along with the audio, by which the media is effectively delivered in a semantically meaningful way to the user, without much loss in pedagogic values. For the key-frame selection from a talking sequence of the instructor, we use a no-reference visual quality metric propounded in [154]. This is because there is nothing available other than the perceptual visual quality of the frame to define the key-frame of such sequences. The key-frames for content segments in the video are extracted in a novel way based on treating *ink pixels* (pixels corresponding to the scribbling by the instructor) as a measure of semantic content to maximize the content dissemination. The horizontal projection profile (HPP) converts the spatially distributed ink pixels in the content frames to useful information for processing the content to extract the keyframe. Since the text regions appear almost horizontally in document images, HPP is found to be a powerful tool for processing it. For the content re-creation through key-frames, one needs the information about the temporal duration for which a particular key frame is a representative frame for the visual content. This information of the time stamp for the display of key-frames is generally provided by a meta-data created on summarization. The audio stream is processed separately to yield a compressed/refined audio file which should be played in perfect synchronism with the display of key-frames so as to render a feel of complete video perception. Since the repackaged media (IMP) may not contain any video segment, the compression factor is extremely high, typically in the range 10-50 over and above the MPEG coding scheme of the original video[1].

Next we extend the content re-creation method to display the IMP on mobile phones, which are so popular now-a-days in student community. While adapting the IMP on mobile devices, care should be taken to preserve the legibility of the written text in content frames. Here we suggest a legibility retentive display (LRD) of the instructional media on miniature screens by which the instructional contents are delivered to the students with the fullest possible resolution. This is accomplished by delivering the textual content in the key-frame using a selection window of size equal to that of the mobile screen and controlling the movement of the window by a meta-data created by tracking the written text in the original video. We suggest to use the combination of HPP and vertical projection profile (VPP) of ink pixels in

---

[1] This is with the audio file in the MP3 format which is not suitable for general voice recording. It may be noted that even higher compression could be achieved by employing a proper audio codec. For example, using Samsung voice pen (SVR-S1540) we found that the compression factor is of the order of 1000, since it employs an SVD file format for audio. Since many of these audio codecs are protected intellectual properties, we refrain from using them while discussing the IMP generation process in this monograph. The key idea here is that the IMP player should be able to select its own data format.

the content frame to derive the text tracking meta-data. Here the meta-data contains not only the temporal marking for which the key frames are valid, but also the x-y co-ordinates of the tracked text. During media re-creation on a mobile device, the non-content key-frame is scaled to fit the mobile screen and displayed for the required time duration. For the content key-frames, a selection window of size equal to that of the mobile screen pan across it in accordance with the available meta-data so as to deliver the local visual content in the *region of interest* (ROI) in its fullest resolution to the viewer. The audio is appended in perfect synchronization with the display of these key-frames. This is actually implemented by developing a mobile multimedia player intended to deliver the content in the key frames and the audio according to the meta data.

Media authentication is of great importance in the present scenario of the increased access to the stored digital data. In the field of education, the producers of instructional videos should be benefited by protecting the ownership of the digital data. The illegal use of the media should be prevented effectively in distance education systems. In this respect, we go for digital watermarking of the IMP before its deployment. Since the IMP is comprised of key-frames and audio, both the media watermarking schemes are dealt with in this monograph. The key-frame watermarking is done using a discrete cosine transform (DCT) based spread spectrum scheme in which the blue component of the color image is marked. The choice of blue channel for watermarking is due to the fact that the human visual system (HVS) is claimed to be less sensitive to this band. We employ a scheme in which each watermark bit is embedded in the mid-band DCT coefficients of each 8×8 block of the image. The mid-band is chosen as the embedding region so as to provide additional resistance to lossy compression techniques (high frequencies), while avoiding significant modification (low frequencies) of the cover image. This method is found to be robust enough against many attacks like scaling, smoothing and filtering. The audio stream is watermarked by using the method propounded in [160]. We employ a a multi-segment watermarking scheme based on the features of audio histogram. The watermark is robust against random cropping and time scale modification (TSM) attacks since the features of audio histogram are insensitive to such attacks. To achieve high detection accuracies, the principle of decision fusion is employed.

Finally, in an aim to help the users of distance education to have a fast preview of the contents of a lecture before downloading or even buying the video for the whole course, we produce a lecture video capsule (LVC). The LVC is formed by merging selected video snippets from the original video, according to a highlight definition strategy. For clip selection from non-content segments like a talking sequence of the instructor, we just adopt the no-reference (NR) visual quality metric suggested in [154]. This is due to the fact that mere perceptual visual quality will suffice to create the highlight for the talking sequence since it plays just a role of situational awareness in lecture videos. For content segments, a novel method of defining the highlights is suggested. We look for the high quality content frames, which are *well-written*. For this, a four-component feature model, based on the statistical parameters of the intensity histogram as well as the horizontal projection profile (HPP) of the ink pixels present in the content frames are used to derive a no-reference objective quality measure. The statistical components measure the mean separation and average sharpness of the text regions compared to its background while the HPP based components measure the pedagogic content and its cleanliness in the frame. The HPP, as applied to ink pixels in a content frame is found to be very effective in analyzing the textual content in video frames. The high quality frame locations are noted for both content and non-content segments in the video to effect the clip selection procedure. We select the video snippets of duration of a few seconds around these good quality frame locations along with the audio and merge with a proportional representation of the detected classes of instructional activity to produce the LVC.

At this point it is better to differentiate between the LVC and IMP. Of course, both are instructional summaries, but what makes it different is the purpose for which they are devised to. An LVC is just a preview video, enabling the students to have a glimpse at the entire course content. Hence it can not be used for learning purpose. Typically a one hour video yields an LVC of a few minutes. Thus on producing an LVC, we map the original video to another video of very short duration. But IMP is never a video file. Instead, it is a collection of key-frames, audio file and a meta-data. Using a programmed playback of all these, the user get a *feel* of the original video. The playback duration is same as that of the original video. It is the meta-data file that provides the necessary linkage between the display of key-frames and the audio playback. Since the visual content in the lecture video is delivered to the student in a video pace using key-frames and the full audio, an IMP can surely be employed for learning purpose.

2.3 Constraints on IMP

We have already realized the fact that the IMP is an efficient representation of its original video in terms of the storage and bandwidth requirements. It is mainly due to the fact that it does not contain any temporal redundancies in the non-content as well as content regions in the video. The redundancy elimination and the meta-data creation are two closely related tasks in the production of IMP. One can say that the meta-data is created on account of eliminating the redundancy.

Since the information in the key-frames are augmented by the audio, the clarity of the audio signal is crucial in IMP. The explanation about the textual regions in the content segments in the video appears in the audio track and so it should be delivered to the viewer with adequate quality. Moreover, the IMP should be able to provide a full lecture to the viewer as in the original video and so the original playback duration of the instructional video should be preserved in the content re-creation through IMP. That is, if the original video is of duration of 1 hour, the resulting IMP should also yield a media playback duration of 1 hour. This is achieved through the necessary time stamps kept in the meta-data and the associated complete audio for 1 hour. In order to have a further perfection, the IMP should contain all the topics covered in the original lecture. For this, the requirement is the accuracy in key-frame selection, in addition to the normally required distinctness of selection. That is, the key-frame selection algorithm should select *all* the distinct key-frames which correspond to the different topics covered in the lecture. Of course, the duplication or redundancy in the key-frames should be avoided.

When the IMP is to be displayed on the small screens of miniature devices, the legibility of the textual content is another issue to be addressed. As already mentioned in the previous section, we solve this problem by a moving key-hole image display of key-frames through which a legibility retentive display of the content is achieved on miniature mobile screens. This requires that the key-hole video is animated on the client device at a sufficiently high frame rate so that the movement of the window appears to be a smooth one.

2.4 Video processing needs

Since our objective is to display a key-frame for the required temporal duration, the original video frame would suffer from resolution when displayed as a static image. This necessitates the enhancement of key-frames, especially content ones, before their static display. An image super-resolution framework is employed for this. Super-resolution is the technique by which we achieve a higher spatial resolution for an image frame from a number of temporally adjoining low-resolution video frames. We treat the sequence of video frames adjoining a particular content key-frame as different low resolution representations of a single high-resolution key-frame image of the instructional scene and use the displacement cues in this low-resolution sequence to achieve super-resolution. We suggest to use a low-resolution sequence of about 10 to 20 frames in the raw video around the location of the selected key-frame to get a content magnification factor of 2 to 4.

It may be noted that there is a possibility of selecting the same frame more than once as a key-frame if the handwritten slides suffer from skew. This is because we employ a horizontal projection profile (HPP) based method for key-frame selection of content frames which naturally assumes that the text portions appear strictly in horizontal direction. However there could be occasional tilt of the writing page as the instructor writes on the paper and so the HPPs of the resulting content frames remain no longer similar and the corresponding frames might be selected as different key-frames. This is not a problem in the case of slide show frames, but could be quite severe in the case of handwritten slides. Hence in order to provide the distinctness for the key-frames, a proper skew correction mechanism should be employed. We use a Radon transform based skew correction by which duplicate key-frames are eliminated.

In order to display the lecture video content on a miniature screen, we have to adopt a retargeting approach by which a selective portion of the content key-frame appears on the small screen with the fullest possible resolution as mentioned earlier. This is called a legibility retentive display in which a content key-frame is displayed through a moving key-hole image riding over it. This moving key-hole image actually selects the *region of interest* (ROI) from the original key-frame. The movement of this key-hole image is made automatic with the tracking co-ordinates contained by the meta-data file. Note that the non-content frames like the talking head of the instructor may just be scaled down and displayed in a conventional manner on the screen since they are assumed to be merely for situational awareness.

2.5 Media repackaging needs

The content re-creation at the client side is through a programmed playback of the IMP by which an estimate of the original lecture is rendered to the end user (student). A crucial factor that should be mentioned here is the synchronization of the audio playback with the display of key-frames. This should be strictly maintained during the media re-creation process, otherwise there would be a semantic mismatch. That is, the instructor's speech which corresponds to his/her talking sequence should remain synchronized on media re-creation. In the same manner, the audio corresponding to the content segments (handwritten/electronic slides) should retain the correspondence on repackaging. The meta-data file might contain the required synchronization information for the audio data, especially if we split the complete audio into several smaller files during repackaging.

When the repackaging is meant for miniature devices which inherently possess small displays, we know that the video meta-data file additionally contains another field called the tracking co-ordinates. This is meant for the selection of the ROI from the key-frames and for rendering only that portion on the screen which would effect a moving key-hole image display of the key-frame. In order to extract these co-ordinates, the task required to be performed is the tracking of the pen or the currently written text. A rectangular area around this co-ordinate having a size comparable with the miniature screen size is referred to as the ROI.

The playback mechanism by which the repackaged media (IMP) is delivered to the viewer should ideally be platform or software independent so that the application is not restricted. Moreover, the IMP player should be an open source material. However, the content (i.e., IMP data) is typically owned by the service provider and hence an appropriate copyright protection mechanism must be enforced. In this monograph, we suggest the use of watermarking for content protection. The IMP player should be able to play the content in presence of the added watermark.

Chapter 3
Current State of the Art

Introduction

Past research efforts in subject areas related to this book are discussed in this chapter. An extensive study of the literature is performed which forms the guidelines for developing the methodology outlined in this book. Broad classifications of subject areas related to this research are video shot detection and recognition, techniques for temporal summarization, techniques for content analysis, content re-creation, content authentication, and video repackaging for miniature devices. The related background research articles collectively contribute to this research on content re-creation of educational video in an attempt for the technology enhancement in distance education.

There are a plenty of literature in the last decade on content based structuring of commercial video for easy access and fast browsing. In [150] authors aim at establishing a large, on-line digital video library featuring full-content and knowledge-based search and retrieval under the Carnegie Mellon's Informedia Digital Video Library project. The project's approach applies several techniques for content-based searching and video sequence retrieval. Content is conveyed in both the narrative (speech and language) and the image forms. The authors have focused their work on two corpora - one is science documentaries and lectures and the other is broadcast news with partial closed-captions. Literature specifically in the educational video domain which aim at enhancing distance education is reported later in the last decade. Authors in [24] present a web-based synchronized multimedia lecture system for distance education. This is designed by streaming video clips for the audio/video lecturing and by dynamically loading HTML pages to present the lecture note navigation process. More literature in this field evolved in the early stages of this decade. A complete procedure for the production and delivery of high amount of information like video and audio data for distance education systems is presented in [35]. Issues related to digital video coding are also addressed along with a special module for the lecture prepared for the deaf people. The focus is on lecture video recording, transmission, digital mounting, encoding, video on demand, and streaming. No efforts have been made for content analysis or repackaging of instructional video by which a much better representation of the original video could be achieved.

A method for structuring lecture videos is presented in [94] which supports both topic indexing and semantic querying of multimedia documents for distance learning applications. Here the authors aim at linking the discussion topics extracted from the electronic slides with the associated video and audio segments. Two major techniques in this approach include video text analysis and speech recognition. A video is first partitioned into shots based on slide transitions, then for each shot, the embedded video texts are detected, reconstructed and segmented as high-resolution foreground texts for commercial optical character recognition (OCR). The recognized texts are matched with their associated slides for video indexing. Meanwhile, phrases (title) and keywords (content) are both extracted from the electronic slides to spot the corresponding speech signal. The spotted phrases and keywords are further utilized as queries to retrieve the most similar slide for speech indexing. In [34] authors try to extract the structure in educational and training media based on the type of material that is presented during lectures and training sessions. The narrative structure that evolves by using different types of presentation means such as slides, web pages, and white board writings is used for segmenting an educational video for easy content access and nonlinear browsing of the material presented. Automatically detecting sections of videos as delineated by the use of supplementary teaching/instructional visual aids allows a proper representation of educational video with high level of semantics, and provides a concise means for organizing educational content according to the needs of different users in e-learning scenarios.

Authors in [85] review previous research work on capturing, analyzing, indexing, and retrieval of instructional videos, and introduce on-going research efforts related to instructional videos. They compare an instructional video with other video genres and address special issues and difficulties in content-based indexing and retrieval of such a video. A content-aware streaming method of lecture videos over wireless networks was proposed in [82] for e-learning applications. Here the authors provide a method for real-time analysis of instructional videos to detect video content regions and classify video frames followed by dynamically compressing the video streams by a *leaky video buffer* model. The adaptive feedback control scheme in the system that transmits the compressed video streams to clients is not only based on the wireless network bandwidth, but also based on the video content and the feedback from video clients. In [49] authors present evolving technologies that facilitate the most important advances in the field of distance education and reviews the existing applications in the scenario of increased popularity of Internet and the World Wide Web. They provide an overview of advanced technologies that facilitate experience-based learning at a distance, reduce the cost of distance education, increase access to education and integrate formal education into the fabric of everyday life.

Having discussed some of the relevant literature on a general purpose e-learning system, we now focus on various video analysis techniques relevant to the e-learning. The different sections in this chapter are organized by categorizing the related literature to different subject areas concerned with the content re-creation aspect of an instructional video with focus on repackaging it on client devices in an e-learning environment. Section 3.1 reviews the methods for video shot detection and recognition. The literature for temporal video summarization is discussed in Section 3.2. Various approaches for content analysis in video are dealt with in Section 3.3 and the literature for content re-creation is presented in Section 3.4. The content (image and audio) authentication methods have been discussed in Section 3.5 and the existing methods for video repackaging for miniature devices are discussed in Section 3.6.

3.1 Video shot detection and recognition

Video shot detection methods in the literature mostly involve heuristics and fail to perform satisfactorily under varied shot detection scenarios. Though model based shot recognition methods are popular, they are inadequate when a given test video sequence contains transitions. The existing shot detection techniques can be classified into two categories : (a) threshold based methods and (b) machine learning based methods.

In [104] authors devise mechanisms for the segmentation of a video into its constituent shots and their subsequent characterization in terms of content and camera work. For shot detection, they suggest a scheme consisting of comparing intensity, row, and column histograms of successive I-frames of MPEG video using the chi-square test. For characterization of segmented shots, they address the problem of classifying the camera motion into different categories using a set of features derived from motion vectors of P and B frames of MPEG video. The central component of the proposed camera motion characterization scheme is a decision tree classifier built through a process of supervised learning. A technique for segmenting a video using hidden Markov models (HMM) is described in [14] for which features include an image-based distance between adjacent video frames, an audio distance based on the acoustic difference in intervals just before and after the frames, and an estimate of motion between the two frames. In contrast to other works where the features for frame difference and audio difference are used separately for video segmentation, this technique allows features to be combined within the HMM framework. Further, thresholds are not required since the automatically trained HMMs replace the role of thresholds.

A statistical framework for optimal shot boundary detection is formulated in [46] which is based on the minimization of average detection error probability. The required statistical functions are modeled by using a metric based on motion compensation for visual content discontinuities and by using the knowledge about the shot length distribution and visual discontinuity patterns at shot boundaries. The authors base on the studies on statistical activity analysis [28, 123] in motion pictures and infer that the distribution of shot lengths fits to Poisson function [102]. Extending this assumption to video shots, they integrate the Poisson curve to get an *a priori* probability for a shot boundary at a certain shot length. Major advantages of this method are its robustness and the sequence-independent detection performance, as well as the possibility of simultaneously detecting different types of shot boundaries.

In [41] authors present the results of performance evaluation and characterization of a number of shot-change detection methods that use color histograms, block motion matching, or MPEG compressed data. Their study is very relevant since it delivers a single set of algorithms that may be used by researchers for indexing video databases. Authors in [15] show that by minimizing the entropy of the joint distribution, an HMM's internal state machine can be made to organize observed video activity into meaningful states. They work with models of office activities and outdoor traffic scenes. The HMM framework is learnt with principal modes of activity and patterns of activity change. This framework is then adapted to infer the hidden states from extremely ambiguous images, in particular, inferring 3D body orientation from sequences of low-resolution silhouettes. In [45] authors use a combination of the Bayesian model for each type of transition and a filtered frame difference called structural information for video shot detection. A method that combines the intensity and motion information to detect scene changes such as abrupt and gradual changes is proposed in [55]. Two features are chosen as the basic dissimilarity measures, and self and cross validation mechanisms are employed via a static scene test. They also develop a novel, intensity statistics model for detecting gradual scene changes. Authors in [169] present an algorithm for automatic shot detection by using a *flash model* and a *cut model* which deal with the false detection due to flashing lights. A technique for determining the threshold that uses a local window based method combined with a reliability verify process is also developed.

In multimedia-based e-learning systems, there are strong needs for segmenting lecture videos into subject topic-based units in order to organize the videos for browsing and to provide a search capability. Automatic segmentation is highly desired because of the high cost of manual segmentation. While a lot of research has been conducted on topic segmentation of transcribed spoken text, most attempts rely on domain-specific cues and a formal presentation format, and require extensive training; none of these features exists in lecture videos containing unscripted and spontaneous speech. In addition, lecture videos usually have a very few scene changes, which imply that the visual information that most video segmentation methods rely on is not available. Furthermore, even when there are scene changes, they do not match with the topic transitions. Authors in [78] make use of the transcribed speech text extracted from the audio track to segment lecture videos into topics. Their approach utilizes features such as noun phrases and combines multiple content-based and discourse-based features.

Even then the problem of segmenting and recognizing complex activities composed of more than one single activity is an ill-addressed one. An HMM-based approach is presented in [96] which uses a suitable threshold and voting to automatically and effectively segment and recognize complex activities in a video. In [173] authors present a method for video shot boundary detection using independent component analysis (ICA). By projecting video frames from an illumination-invariant raw feature space into a low dimensional ICA subspace, each video frame is represented by a two-dimensional compact feature vector. An iterative clustering algorithm based on adaptive thresholding is developed to detect cuts and gradual transitions simultaneously in the ICA subspace. A machine learning based shot detection approach using HMMs is presented in [172] in which both the color and shape clues are utilized. Its advantages are twofold. First, the temporal characteristics of different shot transitions are exploited and an HMM is constructed for each type of shot transition, including cut and gradual transition. As trained HMMs are used to recognize the shot transition patterns automatically, it does not suffer from any trouble of threshold selection problem. Second, two complementary features, statistical corner change ratio and HSV color histogram difference, are used. The former summarizes the shape well whereas the latter summarizes the appearance.

In [117] authors develop a system for human behavior recognition in video sequences, by modeling it as a stochastic sequence of actions. Actions are described by a feature vector comprising both trajectory information (position and velocity), and a set of local motion descriptors. Action recognition is achieved via a probabilistic search of image feature databases representing previously seen actions. HMMs which encode scene rules are used to smooth sequences of actions. A high-level behavior recognition is achieved by computing the likelihood that a set of predefined HMMs explains the current action sequence. Thus, human actions and behavior are represented using a hierarchy of abstraction: from person-centric actions, to actions with spatio-temporal context, to action sequences and, finally, a general behavior. While the upper levels use Bayesian networks and belief propagation, the lowest level uses non-parametric sampling from a previously learned database of actions. The combined method represents a general framework for human behavior modeling. Methods for detecting shot boundaries in video sequences and for extracting key frames using metrics based on information theory are presented in [19]. The method for shot boundary detection relies on the mutual information (MI) and the joint entropy (JE) between the frames. It can detect cuts, fade-ins and fade-outs. The information theory based measure provides better results because it exploits the inter-frame information in a more compact way than a simple frame subtraction. The method for key frame extraction uses MI as well.

Authors in [167] conduct a formal study of the shot boundary detection problem. First, a general framework of shot boundary detection technique is proposed. Three critical issues, i.e., the representation of visual content, the construction of continuity signal and the classification of continuity values, are identified and formulated in the perspective of pattern recognition. Second, a comprehensive review of the existing approaches is conducted. The representative approaches are categorized and compared according to their roles in the formal framework. Based on the comparison of the existing approaches, an optimal criterion for each module of the framework is discussed, which provides a practical guide for developing novel methods. Third, with all the above issues considered, they present a unified shot boundary detection system based on a graph partition model.

In a review paper [145] authors present a comprehensive survey of efforts in the past couple of decades to address the problems of representation, recognition, and learning of human activities from video and related applications. They discuss the problem at two major levels of complexity: i) *actions* and ii) *activities*. Actions are characterized by simple motion patterns typically executed by a single human. Activities are more complex and involve coordinated actions among a small number of humans. They discuss several approaches and classify them according to their ability to handle varying degrees of complexity as interpreted above. They begin with a discussion of approaches to model the simplest of action classes known as an atomic or primitive action that does not require sophisticated dynamical modeling. Then, methods to model actions with a more complex dynamics are discussed. The discussion then leads naturally to methods for a higher level representation of complex activities. An algorithm to segment an object from a stereo video is presented in [152]. The algorithm combines the strong cues of the disparity map with the frame difference. Disparity maps are first segmented to get the object segmentation template along different disparity layers. Then frame difference extractions in the template regions are made to modify the object edge, thus selecting the moving object precisely. The algorithm is efficient in segmentation when there is overlapping in different disparity motion objects or the background is moving.

To overcome the difficulty of getting prior knowledge of the scene duration, in [22] authors propose a method in which the shots are clustered into groups based only on their visual similarity and a label is assigned to each shot according to the group that it belongs to. Then, a sequence alignment algorithm is applied to detect when the pattern of shot labels changes, providing a final scene segmentation result. Shot similarity is computed based only on visual features, while the ordering of shots is taken into account during sequence alignment. To cluster the shots into groups they propose an improved spectral clustering method that also estimates the number of clusters and employs a fast global k-means algorithm in the clustering stage after the eigenvector computation of the similarity matrix. The same spectral clustering method is applied to extract the key-frames from each shot. A shot-based interest point selection approach for effective and efficient near-duplicate search over a large collection of video shots is proposed in [174]. The basic idea is to eliminate the local descriptors with lower frequencies among the selected video frames from a shot to ensure that the shot representation is compact and discriminative. Specifically, the authors propose an adaptive frame selection strategy, called furthest point Voronoi (FPV), to produce the shot frame set according to the shot content and frame distribution. They describe a strategy to extract the shot interest descriptors from a key-frame with the support of the selected frame set.

In [134] authors describe an approach for object-oriented video segmentation based on motion coherence. Using a tracking process based on adaptively sampled points (namely particles), 2-D motion patterns are identified using an ensemble clustering approach. Particles are clustered to obtain a pixel-wise segmentation in space and time domains. The segmentation result is mapped to a spatio-temporal feature space. Thus, the different constituent parts of the scene that move coherently along the video sequence are mapped to regions in this spatio-temporal space. These regions make the temporal redundancy more explicit, leading to potential gains in video coding applications. The proposed solution is robust and more generic than similar approaches for 2-D video segmentation found in the literature.

3.2 Techniques for temporal summarization

A highlight can be defined as a short duration video clip which possesses some local importance. One has to define some domain based rules for the selection of video snippets for highlight creation in a video. There exist a good amount of literature on the highlight creation of sports video. Here the interesting events always occur with an increased crowd activity by which the audio energy level is boosted up in sports video. Hence audio is an important cue for sports video highlight creation. Also, the visual information of the desired highlight, e.g., scoring a goal or a boundary (home run) hit, varies much from the ambient one in sports video. Hence almost all of the methods for sports video highlight creation use predefined models which make use of visual as well as audio signal. Literature addressing the problem of highlight creation in instructional video is not reported yet.

Authors in [57] propose methods for an efficient representation of video sequences through *mosaics*. They describe two different types of mosaics called the static and the dynamic mosaics that are suitable for different needs and scenarios. They also discuss a series of extensions to these basic mosaics to provide representations at multiple spatial and temporal resolutions and to handle the 3D scene information. The techniques for the basic elements of the mosaic construction process, namely alignment, integration, and residual analysis are described along with several applications of mosaic representations, including video compression, enhancement, enhanced visualization, video indexing and search. Techniques to analyze a video and build a compact pictorial summary for visual presentation is proposed in [165]. In this work, a video sequence is condensed into a few images - each summarizing an incident taking place in a meaningful segment of the video. In particular, the authors present techniques to identify the dominance of the content in subdivisions of a segment based on certain analysis results, select a graphic layout pattern according to the relative dominances, and create a set of video posters each of which is a compact, visually pleasant, and intuitive representation of the story content. The collection of video posters arranged in a temporal order then forms a pictorial summary of the sequence to tell the underlying story. The proposed technique and the compact presentation offer valuable tools for video browsing, query, search, and retrieval in the digital libraries and over the Internet.

A method to extract video highlights automatically, so that the viewing time can be reduced is presented in [119]. Here the authors work on baseball videos and detect highlights using audio-track features alone without relying on expensive-to-compute video-track features. They use a combination of generic sports features and baseball-specific features to obtain the results. They also present details on the relative performance of various learning algorithms, and a probabilistic framework for combining multiple sources of information. In [31] authors present an algorithm for automatic extraction of highlights in sports video using audio, text and visual features. The extracted descriptions and annotations can be used for selective browsing of long duration sports videos. A formal learning mechanism to track video browsing behavior of users is presented in [141]. This information is then used to generate fast video previews. Specifically, user transitions while browsing through videos are modeled as the hidden states of an HMM. The authors estimate the parameters of the HMM using the maximum likelihood estimation for each sample observation sequence of a user interaction with videos. Video previews are then formed from interesting segments of the video automatically inferred from an analysis of the browsing states of viewers. Audio coherence in the previews is maintained by selecting clips spanning complete clauses containing topically significant spoken phrases.

In [21] authors describe a statistical method to detect highlights in a baseball game video. Their approach is based on the observations that (i) most highlights in baseball games are composed of certain types of scene shots and (ii) these scene shots exhibit a special transition context in time. To exploit these two observations, they first build statistical models for each type of scene shots with products of histograms, and then for each type of highlight an HMM is learned to represent the context of transition in the time domain. A probabilistic model is obtained by combining the two, which is used for highlight detection and classification. A fully automatic and computationally efficient framework for analysis and summarization of soccer videos is proposed in [36] which uses cinematic and object-based features. The proposed framework includes some novel low-level processing algorithms, such as dominant color region detection, robust shot boundary detection, and shot classification, as well as some higher level algorithms for goal detection, referee detection, and penalty box detection. The system can output three types of summaries - (a) all slow-motion segments in a game, (b) scoring shots for all goals in a game and (c) slow-motion segments classified according to object-based features. The first two types of summaries are based on cinematic features only for fast processing, while the summaries of the last type contain higher level semantics. The proposed framework is efficient, effective, and robust. It is efficient in the sense that there is no need to compute object-based features when cinematic features are sufficient for the detection of certain events, e.g., scoring goals in a soccer game. It is effective in the sense that the framework can also employ object-based features when needed to increase accuracy (at the expense of more computation).

Authors in [98] present a method for event detection, in both gray-level and feature domains. For the gray-level domain, spatio-temporal templates are created by stacking the individual frames of the video sequence, and the detection is performed on these templates. In order to recognize the motion of features in a video sequence, the spatial locations of the features are modulated in time, thus creating a one-dimensional vector which represents an event in the detection process. Authors present three applications - (a) detection of an object under three-dimensional (3-D) rotations in a video sequence, (b) visual recognition of spoken words and (c) recognition of two-dimensional and 3-D sketched curves. The technique is capable of detecting 3-D curves along viewing directions which substantially differ from those in the training set. The resulting detection algorithm is very fast and can successfully detect events even in very low resolution. Also, it is capable of discriminating the desired event from arbitrary events, and not only from those in a negative training set.

A method for creating dynamic video synopsis is presented in [114] in which most of the activities in the video are condensed by simultaneously showing several actions, even when they originally occurred at different time instants. For example, one can create a *stroboscopic movie*, where multiple dynamic instances of a moving object are played simultaneously. Their work is an extension of the stroboscopic still picture [107]. Previous approaches for video abstraction addressed mostly the temporal redundancy by selecting representative key-frames or time intervals. In dynamic video synopsis the activity is shifted into a significantly shorter period in which the activity is temporally much denser. A methodology for the generation of short and coherent video summaries based on clustering of similar activities is presented in [111]. The method is based on the fact that objects with similar activities are easy to watch simultaneously and outliers can be spotted instantly. Clustered synopsis is also suitable for an efficient creation of ground truth data.

In [75] authors propose a video condensation technique in which information is carved out from the space-time video volume in a novel way. This video condensation process is conceptually simple and is relatively easy to implement. The authors introduce the concept of a video ribbon, inspired by that of seam carving method [129] recently proposed for image resizing. They recursively carve ribbons out by minimizing an activity-aware cost function using dynamic programming. The ribbon model they develop is flexible and permits an easy adjustment of the compromise between temporal condensation ratio and anachronism of events. A new visual attention index (VAI) descriptor based on a visual attention model is suggested in [105] to bridge the semantic gap between low-level descriptors used by computer systems and the high-level concepts perceived by human users. With VAI, we can automatically extract key-frames that are most aligned with a human perception both at the shot and the clip levels. The distinct features of this system are: (a) a novel, dynamic attention-computation technique and feature descriptor (that is, VAI) based on a visual attention mechanism, (b) model fusion based on a motion priority scheme and (c) an adaptive video summarization method that allows for the extraction of the most meaningful key-frames based on the changes of the viewers' focus points. Compared with traditional key-frame extraction, this method is claimed to generate high-level semantically correct key-frames from video clips and can control the key-frame density according to the content variation of the whole clip.

3.3 Techniques for content analysis

The objective of video content analysis is many fold - summarization, indexing, retrieval, media re-creation, etc. We concentrate here on content analysis that is closely related to instructional video. The literature presents many different works on the content analysis of commercial as well as instructional video. In [122] authors use principal component analysis (PCA) to reduce the dimensionality of features of video frames for the purpose of content description. The direct use of all the frames of a video sequence is made practical by the low dimensional description. The PCA representation circumvents or eliminates several of the stumbling blocks

in other existing analysis methods. Authors in [146] state that the knowledge about
the structure of video can be used both as a means to improve the performance
of content analysis and to extract features that convey semantic information about
the content. They introduce statistical models for two important components of this
structure, shot duration and activity, and demonstrate the usefulness of these models
with two practical applications. They first develop a Bayesian formulation for the
shot segmentation problem that is shown to extend the standard thresholding model
in an adaptive and intuitive way, leading to an improved segmentation accuracy.
Then by applying a transformation of the shot duration/activity feature space to a
database of movie clips, they illustrate how the Bayesian model captures semantic
properties of the content. They also suggest ways in which these properties can be
used as a basis for an intuitive content-based access to movie libraries.

With the adaptation of e-learning technology in many universities, lectures are
sometimes distributed online in the form of real-time streaming videos. To improve
the quality of these lecture videos, in [51] authors enhance the chalkboard contents
of the lecture videos by using foreground/background separation and a combination technique. The separation allows foreground content to be normalized and de-
noised, thus enhancing the readability of the chalkboard texts. The overall technique
emphasizes the chalkboard contents while preserving the integrity of the natural
video frame. A content-based approach to summarize instructional videos is presented in [84] which focuses on the dominant scene type of handwritten slides. Here
the authors define semantic content as *ink pixels* and present a low-level retrieval
technique to extract this content from each frame with due consideration of various
occlusion and illumination effects. Key frames in the video genre are redefined as a
set of frames that cover the semantic content, and the temporally fluctuating amount
of visible ink is used to drive a heuristic real-time key frame extraction method. A
rule-based method is also provided to synchronize key frames with audio. A gesture driven approach for video editing is suggested in [151]. This approach is to
automatically detect and synchronize the video content with electronic slides in a
lecture video. The gestures in each synchronized topic (or shot) are then tracked
and recognized continuously. By registering shots and slides and recovering their
transformation, the regions where the gestures take place are detected. Based on
the recognized gestures and their positions, the information in slides are seamlessly
extracted, not only to assist video editing, but also to enhance the quality of the
original lecture video.

In [136] authors propose a joint key-frame extraction and object segmentation
method by constructing a unified feature space for both processes, where key-frame
extraction is formulated as a feature selection process for object segmentation in
the context of a Gaussian mixture model (GMM) based video modeling. Specifically, two divergence-based criteria are introduced for key-frame extraction. One
recommends key-frame extraction that leads to the maximum pair-wise interclass
divergence between GMM components. The other aims at maximizing the marginal
divergence that shows the intra-frame variation of the mean density. The proposed
method can extract representative key-frames for object segmentation and so it is a
unique paradigm for content-based video analysis. A method of analyzing instruc-

tional videos using only audio information is presented in [72]. Specifically, an audio classification scheme is first developed to partition the sound-track of an instructional video into homogeneous audio segments, where each segment has a unique sound type such as speech or music. Then the authors apply a statistical approach to extract classroom discussion scenes in the video by modeling the instructor with a Gaussian mixture model (GMM) and updating it on the fly. Finally, they categorize the obtained discussion scenes into either two-speaker or multi-speaker scenarios using an adaptive mode-based clustering approach.

A general architecture and retrieval method for an educational system is discussed in [116] which is based on a knowledge base of existing recorded lectures. The extraction of meta-data from the multimedia resources is one of the main parts of their work. The recorded lectures are transcribed by a speech recognition software. The speech recognition software generates a time stamp for each word. These resources are divided into clusters, marked by timestamps, so that search engines are able to find the exact position of a particular information inside a course. Finally, a retrieval method is presented that allows users to find *example, explanation, overview, repetition* and *exercise* for a particular word or topic-word, which is useful for the student's learning process and allows an easy navigation of the multimedia database.

Authors in [83] present a robust approach to extracting and summarizing the textual content of instructional videos for handwriting recognition, indexing and retrieval, and other e-learning applications. Content extraction from instructional videos is challenging due to image noise, changes in light condition, camera movements, and unavoidable occlusions by instructors. They develop a probabilistic model to accurately detect black board regions and an adaptive thresholding technique to extract the written chalk pixels on the blackboard. They further compute instructional video key frames by analyzing the fluctuation of the number of chalk pixels. By matching the textual content of video frames using Hausdorff distance, they reduce the content redundancy among the key frames. In another work [26], the authors present an approach for summarizing the visual content in instructional videos using middle-level features. They first develop an algorithm to extract textual contents and figures from instructional videos by statistical modeling and clustering. This algorithm addresses the challenges due to image noise, non-uniformity of the black board regions, camera movements, occlusions, and other challenges in the instructional videos that are recorded in real classrooms. Using the extracted text and figures as the middle level features, they retrieve a set of key-frames that contain most of the visual contents.

An approach for the summarization of visual data (images or video) is proposed in [135] which is based on the optimization of a well-defined similarity measure. Here the problem addressed is the retargeting of image/video data into smaller sizes. A good visual summary should satisfy two properties: (a) it should contain as much as possible visual information from the input data, and (b) it should introduce as few as possible new visual artifacts that were not in the input data (i.e., preserve visual coherence). The authors propose a bi-directional similarity measure which quantitatively captures these two requirements. The problem of summarization/retargeting is posed as an optimization problem based on this bi-directional similarity measure. This approach can be used to address a variety of other problems including automatic cropping, completion and synthesis of visual data, image collage, object removal and photo reshuffling.

3.4 Content re-creation

The increasing variety of commonly used display devices, especially mobile devices, requires adapting visual media to different resolutions and aspect ratios. Video retargeting is the process of transforming an existing video to fit the dimensions of an arbitrary display. It aims at preserving the viewers' perceptual experience by maintaining the information content of important regions in the frame, whilst keeping their aspect ratio. The video retargeting problem is further accentuated by the explosion of availability of video content on the web. Though some literature on the retargeting of commercial video is found to appear towards the end of the last decade, none is found to address the problem of retargeting of instructional videos.

An approach for video representation is described in [5] which is based on frame-to-frame alignment, mosaic construction, and 3D parallax recovery. The basic motivation behind this approach is to enable a rapid access to the contents, while maintaining the data in a form as close to the source as possible. This representation supports a wide variety of applications that involve transmission, storage, visualization, retrieval, analysis, and manipulation of video sequences. In [56] authors investigate the dynamic frame skipping strategy in video transcoding. To speed up the operation, a video transcoder usually reuses the decoded motion vectors to re-encode the video sequences at a lower bit rate. When frame skipping is allowed in a transcoder, those motion vectors cannot be reused because the motion vectors of the current frame is no longer estimated from the immediate past frame. To reduce the computational complexity of motion vector re-estimation, a bilinear interpolation approach is developed. Based on these interpolated motion vectors, the search range can be much reduced. Furthermore, they propose a frame rate control scheme which can dynamically adjust the number of skipped frames according to the accumulated magnitude of the motion vectors. As a result, the decoded sequence can present a much smoother motion.

Authors in [166] propose a fast approach to derive a new MPEG stream from an existing MPEG stream, with half the spatial resolution. For the downsized video, the authors first generate from the original compressed video an improved estimate of the motion vectors. Then they propose a compressed domain approach with data hiding to produce DCT residues using an open-loop method. The computational complexity is significantly lower than a number of previously reported approaches. The issue of adapting video streams to different types of terminals with different capabilities such as screen size, amount of available memory, processing power and the type of network access is discussed in [12]. This functionality is very useful for the universal multimedia access application related to MPEG-7 standardization activities. Two different models for transcoding have been examined: rate reduction and resolution reduction.

A video transcoding method is proposed in [76] which combines the use of different transcoding tools to transcode compressed video bit streams to various target bit rates while maintaining a good subjective/objective video quality. The transcoding tools are selected based on the video content which is characterized by two video content descriptors quantifying spatial and motion activities, respectively. These two content descriptors can be readily computed from the compressed video bit stream without full decompression. In [77] authors propose a DCT domain transcoding (DDT) technique in which they extract partial low-frequency coefficients. They investigate fast algorithms of the DDT based on a fast coefficients extraction scheme. They also show that the proposed method can achieve a significant computation reduction while maintaining a close PSNR performance compared to the DDT.

A framework for on-the-fly video transcoding is presented in [30] which exploits computer vision-based techniques to adapt the web access to the user requirements. The proposed transcoding approach aims at coping both with user bandwidth and resources capabilities, and with user interests in the content of a video. The authors propose an object-based semantic transcoding that, according to user-defined classes of relevance, applies different transcoding techniques to the objects segmented in a scene. Object extraction is provided by on-the-fly video processing, without any manual annotation. Multiple transcoding policies are reviewed and a performance evaluation metric based on the weighted mean square error (and the corresponding PSNR), that takes into account the perceptual user requirements by means of classes of relevance, is defined. In [161] authors outline the technical issues and research results related to video transcoding. They also discuss techniques for reducing the complexity, and techniques for improving the video quality, by exploiting the information extracted from the input video bit stream. An overview of several video transcoding techniques and some of the related research issues is provided in [1]. Here the authors discuss some of the basic concepts of video transcoding, and then review and compare various approaches while highlighting critical research issues. They also propose solutions to some of these research issues, and identify possible research directions.

An algorithm for the retargeting of a commercial video is presented in [158] which is based on local saliency, motion detection and object detectors. It consists of two stages. First, the frame is analyzed to detect the importance of each region in the frame. Then, a transformation shrinks less important regions more than important ones. In [155] authors present a framework for the completion of missing information based on local structures. It poses the task of completion as a global optimization problem with a well-defined objective function and derives a new algorithm to optimize it. Missing values are constrained to form coherent structures with respect to reference examples. They apply this method to space-time completion of large space-time *holes* in video sequences of complex dynamic scenes. The missing portions are filled in by sampling spatio-temporal patches from the available parts of the video, while enforcing a global spatio-temporal consistency among all patches in and around the hole. The consistent completion of parts of a static scene simultaneously with dynamic behaviors leads to realistic looking video sequences and images. Space-time video completion is useful for a variety of tasks including (a) video object removal (of undesired static or dynamic objects) by completing the appropriate static or dynamic background information, (b) correction of missing/corrupted video frames in old movies, (c) modifying a visual story by replacing unwanted elements, and (d) creation of video textures by extending smaller primitives.

A simple operator called *seam carving* is proposed in [129] to support image and video retargeting. A seam is an optimal 1D path of pixels in an image, or a 2D manifold in a video cube, going from top to bottom, or left to right. Optimality is defined by minimizing an energy function that assigns costs to pixels. The authors show that computing a seam reduces to a dynamic programming problem for images and a graph min-cut search for video. They also demonstrate that several image and video operations, such as aspect ratio correction, size change, and object removal, can be recast as a successive operation of the seam carving operators. In [118] authors present a video retargeting method using an improved seam carving operator. Instead of removing 1D seams from 2D images they remove 2D seam manifolds from 3D space-time volumes. To achieve this, they replace the dynamic programming method of seam carving with graph cuts that are suitable for 3D volumes. In the new formulation, a seam is given by a minimal cut in the graph and they show how to construct a graph such that the resulting cut is a valid seam, i.e., the cut is monotonic and connected. In addition, they present a novel energy criterion that improves the visual quality of the retargeted images and videos. The original seam carving operator is focused on removing seams with the least amount of energy, ignoring energy that is introduced into the images and video by applying the operator. To counter this, the new criterion is forward looking in time - removing seams that introduce the least amount of energy into the retargeted result. They show how to encode the improved criterion into graph cuts (for images and video) as well as dynamic programming (for images).

An adaptive image and video retargeting algorithm is presented in [63] which is based on Fourier analysis. Here the authors first divide an input image into several strips using the gradient information so that each strip consists of textures of similar complexities. Then, they scale each strip adaptively according to its importance measure. More specifically, the distortions generated by the scaling procedure are formulated in the frequency domain using the Fourier transform. Then, the objective is to determine the sizes of scaled strips to minimize the sum of distortions, subject to the constraint that the sum of their sizes should equal the size of the target output image. They solve the constrained optimization problem using the Lagrangian multiplier technique. They extend the approach to the retargeting of video sequences also. They again present an image and video retargeting algorithm using an adaptive scaling function in [62]. For this, they first construct an importance map which uses multiple features: gradient, saliency, and motion difference. Then they determine an adaptive scaling function, which represents a scaling factor of each column in the source image. Finally, the target image is constructed with a weighted average filter using those scaling factors. The same method is extended to video retargeting too. In

[130] authors present a comparative overview of the latest research in visual media retargeting. They discuss content-aware approaches which, contrary to traditional scaling and cropping, adapt to the salient information within the image or video and rescale the content while preserving visually important information.

In a patent document [54], inventors retarget videos to a target display for viewing with little geometric distortion or video information loss. Salient regions in video frames are detected by using scale space spatio-temporal information. A desired cropping window is selected using a coarse-to-fine searching strategy and the frames are cropped, matching with the aspect ratio of the target display and resized isotropically to match the size of the target display. Video information loss is a result of either spatial loss due to cropping or resolution loss due to downscaling. In another patent document [59], inventors explain a system and method for resizing a set of images while preserving their important content. The saliency of pixels in the set of images is determined using one or more image features. A small number of pixels, called anchor points, are selected from the set of images by saliency based sampling. The corresponding positions of these anchor points in the set of target images are obtained using pixel mapping. To prevent misordering of pixel mapping, an iterative approach is used to constrain the mapped pixels to be within the boundaries of the target image/video. Based on the mapping of neighboring anchor points, other pixels in the target are in-painted by back-projection and interpolation. The combination of sampling and mapping greatly reduces the computational cost yet leading to a global solution to content aware image/video resizing.

3.5 Content authentication

Multimedia watermarking technology has undergone an extensive development during the last decade. Even if it was evolved as a technique for preserving ownership rights, later it found applications in data monitoring, data tracking and steganography. Watermarking is an integral part of the proposed scheme as the distance education systems badly need the protection of the copyright of the lecture videos, avoiding data piracy. In respect to the copyright protection of the instructional media summaries like key-frames and audio, one has to deal with the techniques for watermarking of image and audio. The related literatures for image and audio watermarking are dealt with in this section.

3.5.1 Image authentication

A watermarking technique to add a code to digital images is presented in [108] which operates in the frequency domain for embedding a pseudo-random sequence of real numbers in a selected set of DCT coefficients. Watermark casting is performed by exploiting the masking characteristics of the human visual system, to ensure watermark invisibility. The embedded sequence is extracted without resorting to the original image, so that the proposed technique represents a major improvement to methods relying on the comparison between the watermarked and the original images. In [29] authors present a secure (tamper-resistant) algorithm for watermarking images, and a methodology for digital watermarking that may be generalized to audio, video, and multimedia data. They advocate that a watermark should be constructed as an independent and identically distributed (*iid*) Gaussian random vector that is imperceptibly inserted in a spread-spectrum-like fashion into the perceptually most significant spectral components of the data. They also state that insertion of a watermark under this regime makes the watermark robust to signal processing operations (such as lossy compression, filtering, digital-analog and analog-digital conversion and requantization) and common geometric transformations (such as cropping, scaling, translation and rotation) provided that the original image is available and that it can be successfully registered against the transformed watermarked image. In these cases, the watermark detector unambiguously identifies the owner. Further, the use of a Gaussian noise, ensures a strong resilience to multiple-document or collusion attacks.

A digital watermarking scheme is proposed in [138] to prevent illegal use of digital documents. Here the authors highlight the basic mechanism of watermarking and explain the properties of spread spectrum and its use in watermarking. They also discuss specific issues in watermarking of text, images, and video along with examples. In [140] authors review developments in transparent data embedding and watermarking for audio, image, and video. They discuss the reliability of data-embedding procedures and their ability to deliver new services such as viewing a movie in a given rated version from a single multicast stream. They also discuss the issues and problems associated with copy and copyright protection and assess the viability of current watermarking algorithms as a means for protecting copyrighted data. Authors in [50] review the requirements and applications for watermarking. Applications include copyright protection, data monitoring, and data tracking. The basic concepts of watermarking systems are outlined and illustrated with proposed watermarking methods for images, video, audio, text documents, and other media. Robustness and security aspects are also discussed in detail. The problem of watermark embedding and optimum detection in color images is addressed in [126] which makes use of spread spectrum techniques both in space (direct sequence spread spectrum or DSSS) and frequency (frequency hopping). It is applied to RGB and other opponent color component representations.

A discrete cosine transform (DCT) domain spread spectrum watermarking technique is proposed in [52] for copyright protection of still digital images. The DCT is applied in blocks of 8×8 pixels, as in the JPEG algorithm. The watermark can encode information to track illegal misuses. For flexibility purposes, the original image is not necessary during the ownership verification process, so it must be modeled as a noise source. Two tests are involved in the ownership verification stage: watermark decoding, in which the message carried by the watermark is extracted, and watermark detection which decides whether a given image contains a watermark generated with a certain key. The authors apply generalized Gaussian distributions to statistically model the DCT coefficients of the original image and show how the resulting detector leads to considerable improvements in performance with respect to the correlation receiver, which has been widely considered in the literature and makes use of the Gaussian noise assumption. A different approach to the watermarking evaluation problem is presented in [106] which splits the evaluation criteria into two (independent) groups: functionality and assurance. The first group represents a set of requirements that can be verified using an agreed series of tests and the second is a set of levels to which each functionality is evaluated. These levels go from zero or a low value to very high value. The authors also investigate how evaluation profiles can be defined for different applications and how importance sampling techniques could be used to evaluate the false alarm rate in an automated way. In [69] authors discuss the need for watermarking and its requirements. They present a study of various digital watermarking techniques that are based on correlation and that are not based on correlation.

Authors in [25] present an overview and summary of recent developments on document image watermarking and data hiding. They also discuss those important issues such as robustness and data hiding capacity of the different techniques. In [109] authors present a general framework for watermark embedding and detection/decoding along with a review of some of the algorithms for different media types described in the literature. They highlight some of the differences based on applications such as copyright protection, authentication, tamper detection, and data hiding as well as differences in technology and system requirements for different media types such as digital images, video, audio and text. An adaptive watermarking scheme for color images is proposed in [80]. To increase the robustness and perceptual invisibility, a model called image complexity and perceptibility model (ICPM) is proposed to analyze the properties of different sub-blocks. By utilizing the properties of the image itself, a technique is proposed for watermark casting and retrieval. A DCT domain watermarking technique is presented in [8] which is expressly designed to exploit the peculiarities of color images. The watermark is hidden within the data by modifying a subset of full-frame DCT coefficients of each color channel. Detection is based on a global correlation measure which is computed by taking into account the information conveyed by the three color channels as well as their interdependency. To ultimately decide whether or not the image contains the watermark, the correlation value is compared to a threshold. With respect to existing gray scale algorithms, a new approach to threshold selection is proposed, which permits reducing the probability of missed detection to a minimum, while ensuring a given false detection probability.

A reversible watermarking algorithm is developed in [4] which possesses a very high data-hiding capacity for color images. The algorithm allows the watermarking process to be reversed, which restores the exact original image. The algorithm hides several bits in the difference expansion of a pair of adjacent pixels. The required reversible integer transform and the necessary conditions to avoid underflow and overflow are derived for any data vector of arbitrary length. Also, the potential payload size that can be embedded into a host image is discussed, and a feedback system for controlling this size is developed. In addition, to maximize the amount of data that can be hidden into an image, the embedding algorithm can be applied recursively across the color components. In [2] authors focus on visually meaningful color image watermarks and construct a new digital watermarking scheme based on the discrete cosine transformation. The proposed method uses the sensitivity of human eyes to adaptively embed a watermark in a color image. In addition, to prevent tampering or unauthorized access, a new watermark permutation function is proposed, which causes a structural noise over the extracted watermark. They have also proposed a procedure to eliminate this noise to decrease false positives and false negatives in the extracted watermark. They show that embedding the color watermark adapted to the original image produces the most imperceptible and the most robust watermarked image under geometric and volumetric attacks.

Authors in [144] present two vector watermarking schemes that are based on the use of complex and quaternion Fourier transforms and demonstrate to embed watermarks into the frequency domain that is consistent with our human visual system. Watermark casting is performed by estimating the just-noticeable distortion of the images, to ensure watermark invisibility. The first method encodes the chromatic content of a color image into the CIE $a^* b^*$ chromaticity coordinates while the achromatic content is encoded as CIE L tristimulus value. Color watermarks (yellow and blue) are embedded in the frequency domain of the chromatic channels by using the spatiochromatic discrete Fourier transform. It first encodes a^* and b^* as complex values, followed by a single discrete Fourier transform. The most interesting characteristic of the scheme is the possibility of performing watermarking in the frequency domain of chromatic components. The second method encodes the $L a^* b^*$ components of color images and watermarks are embedded as vectors in the frequency domain of the channels by using the quaternion Fourier transform. Robustness is achieved by embedding a watermark in the coefficient with positive frequency, which spreads it to all color components in the spatial domain and invisibility is satisfied by modifying the coefficient with negative frequency, such that the combined effects of the two are insensitive to human eyes.

An image authentication technique is proposed in [3] which embeds a binary watermark into a host color image. In this scheme the color image is first transformed from the RGB to the YCbCr color space. The watermark is embedded into the Y channel of the host image by selectively modifying the very low frequency parts of the DCT transformation. It is shown that the proposed technique can resist classical attacks such as JPEG compression, low pass filtering, median filtering, cropping, and geometrical scaling attack. Moreover, the recovery method is blind and does not need the original host image for extraction.

3.5.2 Audio authentication

We now review some of the literature on watermarking for audio data. In [11] authors explore both traditional and novel techniques for addressing the data-hiding process and evaluate these techniques in the light of three applications - copyright protection, tamper-proofing and data embedding. A method to embed a digital signature in multimedia documents like audio, image or video is proposed in [163]. This method is based on bit plane manipulation of the LSB and the decoding is easy due to the property of the signature that the authors used. The proposed technique for digital watermarking is also compatible with JPEG processing and can survive after JPEG encoding. An audio watermarking method in time domain is proposed in [9] which offers copyright protection to an audio signal. The modifications in the strength of audio signal is limited by the necessity to produce an output signal that is perceptually similar to the original one. The watermarking method presented here does not require the use of the original signal for watermark detection. The watermark signal is generated using a key, i.e., a single number known only to the copyright owner. Watermark embedding depends on the audio signal amplitude and frequency in a way that minimizes the audibility of the watermark signal. The embedded watermark is robust to common audio signal manipulations like audio coding, cropping, time shifting, filtering, re-sampling, and requantization.

Authors in [87] introduce an improved spread spectrum technique for watermarking. When compared with the traditional spread spectrum SS), the signal does not act any more as a noise source, leading to significant gains. In some examples, reported performance improvements over SS are 20 dB in signal-to-noise ratio (SNR) or an order of magnitude in the error probability. The proposed method achieves roughly the same noise robustness as quantization index modulation (QIM) but without the amplitude scale sensitivity of QIM. A self-synchronization algorithm for audio watermarking is proposed in [159] to facilitate an assured audio data transmission. The synchronization codes are embedded into audio with the informative data, thus the embedded data have the self-synchronization ability. To achieve robustness, the authors embed the synchronization code and the hidden informative data into the low frequency coefficients in the discrete wavelet transform (DWT) domain. By exploiting the time-frequency localization characteristics of DWT, the computational load in searching the synchronization code has been dramatically reduced, thus resolving the contending requirements between robustness of the hidden data and efficiency of synchronization. The performance of the proposed scheme in terms of signal to noise ratio (SNR) and bit error rate (BER) has also been analyzed. An estimation formula that connects SNR with the embedding capacity has been provided to ensure the transparency of the embedded data. BER under a Gaussian noise corruption has been estimated to evaluate the performance of the proposed scheme.

A content dependent, localized robust audio watermarking scheme is proposed in [71] which combat the attacks like random cropping and time scale modification (TSM). The basic idea is to first select steady, high-energy local regions that represent edges like sound transitions and then embed the watermark in these regions. Such regions are of great importance to the understanding of music and will not be changed much for maintaining a high auditory quality. In this way, the embedded watermark has the potential to escape all kinds of manipulations. In [160] authors present a robust audio watermarking scheme by utilizing the properties of audio histogram which is used in this monograph for the watermarking of the instructional audio in Chapter 7. It relies on the fact that the shape of the audio histogram is insensitive to time scale modification (TSM) and random cropping operation. The authors address the insensitivity property through both mathematical analysis and experimental testing and represent the audio histogram shape by some relations of the number of samples in groups of three neighboring bins. By reassigning the number of samples in these groups of bins, the watermark sequence is successfully embedded. During the embedding process, the histogram is extracted from a selected amplitude range by referring to the mean in such a way that the watermark will be able to be resistant to amplitude scaling and an exhaustive search is avoided during the extraction process. The watermarked audio signal is perceptibly similar to the original one.

3.6 Repackaging for miniature devices

A rapid progress in multimedia signal processing has contributed to the extensive use of multimedia devices with a small liquid crystal display (LCD) panel. With the proliferation of mobile devices having a small display, the video sequences captured for normal viewing on a standard TV or HDTV may give the small-display viewers an uncomfortable experience in understanding what is happening in a scene. For instance, in a soccer video sequence taken by a long-shot camera, the tiny objects (e.g., soccer ball and players) may not be clearly visible on the small LCD panel. Thus, an intelligent display technique is needed for viewers having access to a small display. Literature on retargeting of commercial videos for mobile devices are found to be available while that for instructional videos are not yet in place.

Authors in [37] present an approach which allows users to overcome the display constraints by zooming into video frames while browsing. An automatic approach for detecting the focus regions is introduced to minimize the amount of user interaction. In order to improve the quality of the output stream, a virtual camera control is employed in the system. They also show that this approach is an effective way for video browsing on small displays. In [13] authors present buffer-based strategies for temporal video transcoding for mobile video communication in a real-time context. A method of video retargeting is introduced in [79] which adapts a video to better suit the target display, minimizing the loss of important information. Here the authors define a framework that measures the preservation of the source material, and methods for estimating the important information in the video. Video retargeting crops each frame and scales it to fit the target display. An optimization process minimizes the information loss by balancing the loss of details due to scaling with the loss of contents and composition due to cropping. The cropping window can be moved during a shot to introduce virtual pans and cuts, subject to constraints that ensure a cinematic plausibility.

A fully automatic and computationally efficient method for intelligent display of soccer video on small multimedia capable mobile devices is proposed in [128]. In this, instead of taking visual saliency into account, the authors adopt a domain-specific approach to exploit the attributes of the soccer video. The proposed scheme includes three key steps: ground color learning, shot classification, and *region of interest* (ROI) determination. In [53] authors present the use of recorded lectures in formats for mobile devices (video iPod and PDA) in a graduate computer science program using a software system developed at Kennesaw State University, Georgia. Recordings are converted to formats for mobile devices along with an interface that allows students to subscribe to a course to receive automatic lecture downloads. The technology, pedagogy, student perception of the technology, and the impact on educational success are also explored. The lectures through the mobile device are in addition to the existing distance-learning option that allows students to attend live lectures by using a web browser, and to view recorded lectures. Students are able to self-select whether to attend an on-campus classroom, attend live over the Internet, or to view the recorded lectures.

A multimedia gateway implementation which provides video adaptation to mobile clients is described in [16] which uses a multidimensional compressed domain transcoding mechanism. Authors in [139] give an overview of their group's recent and ongoing work on creating a flexible, intuitive, and powerful interface to improve video browsing on hand-held devices in a way similar to how the iPhone's interaction techniques revolutionized mobile navigation of static media. They also present several developed and evaluated concepts and related user-interface designs. A multidimensional transcoding technique is employed in [17] for adaptive video streaming. This method is developed as a solution to the problem of limited capacity of mobile devices, in terms of resolution, power and bandwidth. In [157] authors address the issues of accessing a lecture video content through mobile devices by manipulating two key elements of the lecture video: displayed slides and laser pointer gestures. Displayed slides need to be very crisp compared to the background content. As the content is available from the presentation slides, they describe a method for splicing them into the video on a client site, increasing fidelity and reducing bandwidth needs. The operation removes laser pointer gestures which are often lost due to compression, and are hard to see on the small screens of the mobile display unit. But these gestures are part of what makes watching a lecture video different from simply looking at the slides. Hence they interpret the laser pointer gestures as they analyze the videos, creating representations that can be transmitted at a low cost. These representations can then be iconified on the client side and displayed clearly.

In a patent document [40], inventors disclose a method for intelligently creating, consuming and sharing video content on mobile devices. In this, a video capture mechanism inputs the video data, a key-frame extractor automatically selects at least one key-frame from the video data and a video encoder encodes the video with a key-frame identifier. The key-frame extraction is based on statistical as well as semantic features of video frames. The low-level features include color histograms and motion vectors and the semantic labels include face/non-face, indoor/outdoor, etc. The audio stream also undergoes similar levels (low and semantic) of processing to assist the key-frame selection. The meta-data associated with a specific key-frame is used for video encoding. A device preference module in the system can fix the

number of key-frames preferred by the user, depending on the processing power of
the device, available memory and other device specifications.

A novel method for adapting lecture videos on mobile phones is proposed in
[92] which aims at viewing those videos on low bandwidth connections at a low
cost. The idea is to exploit the redundancy present in lecture videos, such as non-
changing presentation slides. The authors define study elements within the video
that have varying levels of redundancy and user expectations. They transmit images
at different intervals, such as one image every five seconds, according to the study
element involved along with the continuous audio. In [97] authors develop methods
for the adjustment of video in standard definition resolution to the smaller resolution
used in mobile devices. They focus on a region of interest (ROI) based adaptation of
news broadcasting and interview scenes, so that the ROI contains the speaker's face.
At the beginning of every scene, the ROI is marked by an editor, followed by an algorithm for tracking the ROI. The algorithm estimates the global motion caused by
the camera and the local motion caused by the speaker's movements. They estimate
the motion parameters using horizontal and vertical projections. The projection slice
theorem and the Mellin transform are used for reducing the computational complexity.

A low complexity content-aware video retargeting method is proposed in [93]
which is suitable for mobile multimedia devices. Here the authors propose two
types of energy functions for intra- and inter-frames to obtain the significance of
each pixel in the video frames. Since the energy functions are defined using only
the compressed-domain information such as the discrete cosine transform (DCT)
coefficients and motion vectors (MVs), instead of a fully decoded pixel domain
information, the spatio-temporal energy map is constructed without excessive computations. Based on the proposed energy function, they also introduce a fast seam
carving method for mobile devices that carves out the one-dimensional (1-D) seams
without introducing annoying artifacts such as jitter and flicker, taking into account
the object motion based on MVs. Authors in [73] present a multi-scale trajectory optimization based retargeting method for commercial videos on mobile devices. They
formulate video retargeting as the problem of finding an optimal trajectory for a
cropping window to go through the video, capturing the most salient region to scale
towards a proper display on the target. To measure the visual importance of every
pixel, they utilize the local spatial-temporal (ST) saliency and face detection results.
The spatio-temporal movement of the cropping window is modeled using a graph-
based method such that a smoothed trajectory is resolved by a max-flow/min-cut
method. Based on the horizontal/vertical projections and the graph-based method,
the trajectory estimation of each shot can be conducted within one second. Also, the
process of merging trajectories is employed to capture a higher level of saliency in
video.

The successive chapters introduce the various methods adopted for the content
re-creation of instructional videos for distance education purpose. These methods
include repackaging of lecture video to yield an *instructional media package* (IMP),
a legibility retentive display of instructional media on mobile devices, watermarking
of the repackaged media, and the generation of a *lecture video capsule* (LVC). These

approaches largely rely on the fact that instructional videos have a relatively less scene complexity and motion. Hence for the content analysis of the instructional video we adopt a method that is built on a shot segmented video. The next chapter introduces the preliminary procedures that should be done before performing higher level analyses of an instructional video.

Chapter 4
Preprocessing of Lecture Videos

4.1 Introduction

Since a lecture video is shot in a live classroom or at the studio of distance education centers, its scene complexity and motion are found to be very low, which is already discussed in section 1.4. One may note that there are only a few classes of activities (shots) present - (1) an instructor talking about some subject matter and the camera dwelling on him/her, (2) his/her hand writing on a piece of paper or board, (3) showing electronic slides, (4) video imported from other sources for explanation or demonstration purpose, and (5) the audience (students) asking some doubts occasionally. Since the instructor continuously speak when he writes or shows the slides, one may insert his talking face in the empty regions of the handwritten or electronic slide during editing which gives rise to picture-in-picture (PIP) scenes. Fig. 4.1 illustrates these scene types. Note that the scene types in Fig. 4.1 (a), (b), and (c) would be having a very high possibility of occurrence compared to those in Fig. 4.1 (d), (e), and (f). Hence the prime concern is to process the first three scene types which are prominent in lecture videos. In this regard we follow the following terminology throughout this book. Those video frames in which the instructor's face appears are termed as *talking head* (TH) frames, those in which the instructor writes with his hand are termed as *writing hand* (WH) frames, and those in which electronic slides are shown are termed as *slide show* (SS) frames. Hence the preliminary steps before performing any higher level analysis of the instructional video should aim at detecting the shot breaks and classifying the type of scenes found in-between. These two basic steps are called *temporal segmentation* and *activity recognition*. Sometimes the methods for shot detection and recognition are found to be combined in a single algorithm to achieve both goals simultaneously.

Temporal segmentation of a video into its constituent shots is the basic step towards the exploration about the organization of digital video for all higher level analysis. The commonly found shot transitions are (a) hard cuts, (b) fades, (c) dissolves and (d) wipes. We have already mentioned in section 1.4 that a lecture video is subjected to less camera work and editing on its production. Hence most of the

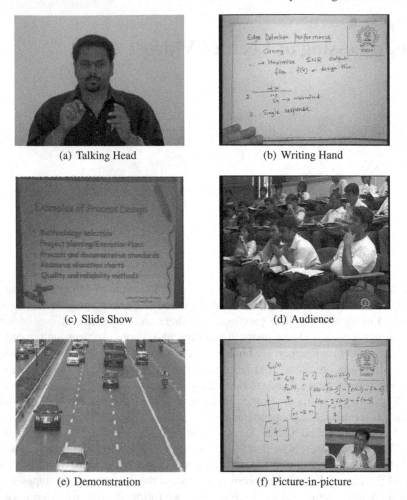

(a) Talking Head

(b) Writing Hand

(c) Slide Show

(d) Audience

(e) Demonstration

(f) Picture-in-picture

Fig. 4.1 Illustration of the possible types of scenes in lecture videos. Note that (a - c) are prominent but (d - f) rarely occur.

scene transitions found in the case of lecture video are hard cuts. Occasionally one can also find scene dissolves. Typical examples of transitions found in lecture videos are shown in Fig. 4.2. In Fig. 4.2(a), the scene transition is a hard cut from TH to WH, in (b) it is a dissolve from TH to WH, in (c) it is a dissolve from TH to SS, and in (d) it is a dissolve from SS to TH. Note that all the dissolves are over a duration of eight frames. These transition examples are taken from different instances of the same video. Since the performance of scene change detection has a direct impact on the subsequent higher level video analysis, a reliable method should be adopted for temporal segmentation [47]. Scene present in between the boundaries are then to be classified into the defined classes of instructional activities.

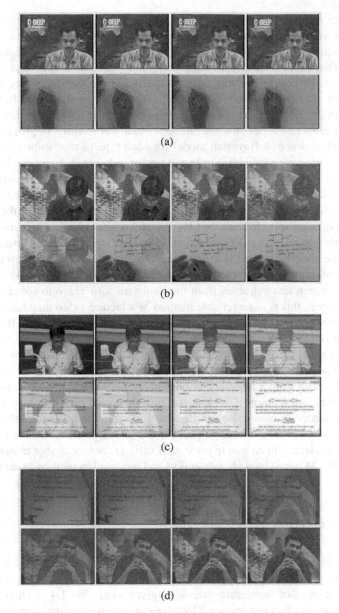

Fig. 4.2 Examples of shot transitions. (a) shows a hard cut transition from TH to WH, (b) shows a dissolve transition from TH to WH, (c) shows a dissolve transition from TH to SS, and (d) shows a dissolve transition from SS to TH.

The existing shot detection techniques can be classified into two categories : (a) threshold based methods and (b) machine learning based methods, where the former usually employs some function of frame difference for pixels, blocks or histograms

[168], which relies on a suitable threshold and the latter employs machine learning approaches which avoids the difficulties involved in threshold selection. Machine learning methods exploit the statistical characteristics of the test data and uses a hidden Markov model (HMM) for classification. In [14] the authors use the differences in signal in both audio and video channels to train the HMM and hence the method does not work well when the scene cuts are not hard enough. In [172] the authors train the HMM using features derived during the transition phases. Hence they cannot handle all types of scene changes equally efficiently. In [45] the authors use the combination of a Bayesian model for each type of transition and a filtered frame difference called structural information for video shot detection. In [66] the authors employ HMM for parsing news video for simultaneous segmentation and characterization. Since it considers audio features along with visual features, it is not suitable for shot detection in the case of lecture videos in which the audio is a fairly continuous speech track. In [46] the authors formulate a statistical model for shot detection, using a metric based on motion compensation for visual content discontinuities and by using the knowledge about the shot length distribution and visual discontinuity patterns at shot boundaries. In [78] the authors make use of the transcribed speech text extracted from the audio track of video to segment lecture videos. However, this is not a reliable method as a lecture video may have different types of (audio) pauses, introducing unnecessary cuts. So depending on the peculiar scene composition of lecture video, one has to adopt simple but reliable methods of shot detection and recognition. Next section deals with a fast method of histogram based video segmentation followed by a method of HMM based activity recognition in section 4.3. The temporal segmentation is meant for identifying the change points and the activity recognition is for categorizing the video activity present between two shot boundary frames. It is seen that the histogram based temporal segmentation shows an ambiguity in change point if the shot transition involved are other than hard cuts. Hence in an aim to cope with varied scenario of shot transitions, we discuss an HMM based joint shot detection and recognition method in section 4.4.

4.2 Shot detection

This approach is developed under the assumption that an instructional video does not contain much motion or any complex type of scene transition. It is seen that this holds for most of the lecture videos, as given in section 1.4 of chapter 1. The objective here is to find out temporal locations at which the visual content in video frames exhibits a nature of transition. These transitions have already been dealt with in the previous section. We aim at a fast shot change detection on simple transitions like hard cuts. To achieve this goal, some quantitative measure of the visual content in these video frames are to be employed. The histogram is found to be very useful in this respect. The histograms of two consecutive frames in a video will be very close to each other if they belong to the same scene, under the assumption that the motion is very low. Wherever there is a scene change, histograms tend to deviate from each other. Hence we use the histogram difference (HD) measure [170] for the temporal segmentation of lecture video. It is given by

$$D(t) = \sum_{i=0}^{255} \mid h_t(i) - h_{t-1}(i) \mid \tag{4.1}$$

where h_t is the normalized intensity histogram of the current frame and h_{t-1} is that of the previous frame. If the sum of the absolute difference of the histograms, expressed in equation (4.1) crosses a threshold, the corresponding frame is declared as a shot-boundary frame. This histogram difference method works well, as there is very little camera movement and hence, a simple but fast and quite accurate scene change detection can be performed.

4.3 Shot recognition

Since the key-frames would represent the content and non-content segments of the video in the schemes proposed in the following chapters, these segments are to be identified first. Since the HMM [112, 113] is a powerful tool for activity recognition, it is used in our approach. Several works have been reported on HMM based activity detection in video [15, 96, 117]. We adopt a method of using a well trained HMM with just two features extracted from the frames of the lecture video sequence. This method has been demonstrated to be quite suitable for scenes that can be modeled by appropriate HMMs, like commercials, news video, sports video, instructional (educational) video, etc.

Hidden Markov models are stochastic state transit models which make it possible to deal with time-scale variability in sequential data [112]. The basic characteristic of HHM is their learning ability which automatically optimizes the parameters of the model when presented with a time sequential data. HMM consists of a fixed, known number of states. Each of these states is assigned a probability of transition from that particular state to any other state including itself. At every instant of time, a transition from one state to another occur stochastically and similar to Markov models, a state at a point in time depends only on the state at the preceding time. Each state yields a symbol according to the probability distribution assigned to the state. The present state or the sequence of states are not directly observable, hence the name hidden Markov model. These states can be inferred through the sequence of the observed symbols. In [112] the author formulated the observations in each state as weighted mixtures of any log-concave or elliptically symmetric probability density functions with mean vector μ_{jm}, and covariance matrix $\mathbf{U}_{jm}$, for the mth mixture component in state j. Gaussian mixtures were used to model the probability distributions of the observations. The weights in the mixture should sum up to one.

The HMM based activity recognition has two phases, namely the training phase and the testing phase. The frame sequence $I = \{I_1, I_2, ..., I_T\}$ is transformed into observation sequences $X = \{X_1, X_2, ..., X_T\}$ for the learning and recognition phases of HMM, where X_n is the feature vector. In the training phase, for each class of instructional activity, a feature vector X_t is extracted from each frame I_t. We consider only the three prominent classes of activities in this book: (1) talking head (TH), (2) writing hand (WH) and (3) slide show (SS). These classes have already been illustrated in Fig. 4.1 (a - c). Since the talking head sequence contains the instructor's face meant for the situational awareness, it is treated as *non-content* frames meaning that there is no instructional content. The writing hand and slide show sequences are treated as *content* frames since they are rich with instructional content. Of course, the audio sequence is always having a high instructional content, irrespective of the associated frame sequence.

One can note that the visual information contained by the first class (TH) differ considerably from the second (WH) and third (SS). The latter two classes share some textual content in common, but differ only in their method of creation. Also, in WH there is a possibility of an occluding hand but SS would be almost free from any occlusion. This is from a single frame perspective, if we come across frames there is another important cue namely *motion*. Hence we select features by considering both these inter-frame and intra-frame aspects. Note that motion in TH is more, compared to that in WH which in turn, is greater than that in SS. Hence the energy of the temporal derivative in intensity space can be used as a relevant feature, which is given by

$$x_1(t) = \sum_{m=1}^{M} \sum_{n=1}^{N} (I_t(m,n) - I_{t-1}(m,n))^2 \qquad (4.2)$$

where I_t is the pixel intensity values of the current frame and I_{t-1} is those of the previous frame, M is the number of rows and N is the number of columns in the video frame.

The gray-level histogram gives the distribution of image pixels over different intensity values. The histogram will be very sparse for the slide show class and moderately sparse for the writing hand and dense for the talking head. Hence the entropy of the histogram [101, 100, 81] can be treated as another good feature for the detection of these activities, which is given by

$$x_2(t) = -\sum_{i=0}^{255} h_t(i) log(h_t(i)). \qquad (4.3)$$

Since the scenes associated with the subject videos is less complex as already explained, these features are found to be adequate for good recognition accuracy. Limiting the features to two is also justified by the fastness criteria. Hence two features which are relevant in the context of instructional scenes are defined, yielding the feature vector, $X_t = \{x_1(t), x_2(t)\}$. An HMM with a Gaussian mixture model with two states is assumed for each class of the instructional activity. Each model is to be learnt by properly chosen training sequences before they are employed for recognition purposes.

4.3.1 Learning the activities

To apply HMMs to time-sequential data from the images $\mathbf{I} = \{I_1, I_2, \ldots, I_T\}$, the images must be transformed into observation sequences $\mathbf{X} = \{X_1, X_2, \ldots, X_T\}$ during the learning and the recognition phases where $X_n = \{x_{1,n}, x_{2,n}, \ldots, x_{J,n}\}$ and J is the dimensionality of the features. For the given example, $J = 2$. In the learning phase, for every class of activity, from each frame I_i of an image sequence, a feature vector $f_i \in \mathbb{R}^n$, is extracted, and the probability density function is constructed. Training an HMM means estimating the model parameters $\lambda = (A, B, \pi)$ for a given activity by maximizing the probability of the observation sequence $\Pr(X|\lambda)$, where A is the state transition probability matrix containing the probability of transition from one state to another, B is the observation symbol probability matrix, consisting of the probability of observing a particular symbol in any given state and π is the initial state probability [112]. The Baum-Welch algorithm is used for estimating these parameters [112]. For each and every activity we calculate the features from the sequence of images. These features of $m = 1, 2, \ldots, M$ datasets of a single type of activity are then stacked together to get a three dimensional data matrix, $\mathbf{X}_{tm}^j$, with other two dimensions being the time n and feature dimension j. This is then fed into Baum-Welch algorithm to learn the optimized set of parameters λ for every activity.

For example, the training feature set from ten video frames for the three classes TH, WH and SS are given below.

$$X_{TH} = \begin{bmatrix} 22.1 & 18.2 & 17.8 & 21.6 & 15.9 & 34.7 & 20.0 & 12.5 & 10.3 & 17.9 \\ 5146 & 5147 & 5148 & 5148 & 5149 & 5149 & 5144 & 5148 & 5147 & 5147 \end{bmatrix}$$

$$X_{WH} = \begin{bmatrix} 15.6 & 17.8 & 15.6 & 23.9 & 18.6 & 18.2 & 13.9 & 16.7 & 16.2 & 17.1 \\ 4621 & 4616 & 4608 & 4605 & 4603 & 4601 & 4610 & 4613 & 4610 & 4614 \end{bmatrix}$$

$$X_{SS} = \begin{bmatrix} 13.8 & 16.1 & 14.1 & 17.9 & 16.5 & 14.7 & 13.5 & 15.1 & 13.8 & 12.6 \\ 3766 & 3768 & 3756 & 3761 & 3754 & 3756 & 3758 & 3758 & 3758 & 3748 \end{bmatrix}$$

Their corresponding state transition probability matrices as estimated are

$$A_{TH} = \begin{bmatrix} 0 & 1 \\ 0.3333 & 0.6667 \end{bmatrix}$$

$$A_{WH} = \begin{bmatrix} 0.6667 & 0.3333 \\ 0.1667 & 0.8333 \end{bmatrix}$$

$$A_{SS} = \begin{bmatrix} 0.5000 & 0.5000 \\ 0.2000 & 0.8000 \end{bmatrix}$$

4.3.2 Recognizing the activities

In the recognition phase, from each frame of the image sequence of the test data, feature vectors are extracted in a similar manner. These vectors are compared against the models for each class of activity . For a classifier of C categories, we choose the model which best matches the observations from C HMMs $\lambda_i = \{A_i, B_i, \pi_i\}$, $i = 1, \dots, C$. This means that when a sequence from an unknown category is given, we calculate the likelihood that this particular observation was from any of the category i for each HMM and select λ_{c^*}, where c^* is the category with the best match. The class that gives the highest likelihood score calculated using the forward algorithm is declared as the winner [112]. Although the HMM is a well appreciated method for activity recognition, the key factor is the choice of the feature vector. We have already explained the selection of features based on the peculiarity of the scenes present in lecture videos in section 4.3. Since it is easy to select two relevant features associated with the visual variation in the constituent scenes in the video, we achieve an excellent recognition accuracy. This is also well supported by the fact that the scene complexity is quite low for lecture videos.

This shot recognition method works under the assumption that the shot boundaries are known. Thus the temporal segmentation step is a prerequisite of this method. Since the histogram based temporal segmentation fail to perform well for gradual or other complex types of shot transitions, we have to devise a method which is able to detect the shot changes in a video, irrespective of the nature of transition and to continuously classify the shots in between. Such a method is explained in next section.

4.4 Joint shot detection and recognition

The shot detection and recognition methods explained in the previous sections hold good only for simple scene transitions like hard cuts. In order to effectively solves the problem of shot detection and recognition of videos involving complex scene transitions we formulate this joint method. Note that most of the video shot detection methods found in the literature involves some heuristics and thus fail to perform satisfactorily under varied shot detection scenarios. Model based methods are also inadequate when a given test video sequence contains transitions. Under these circumstances, a shot detection method which deal with the changes in activities in areas where one has to recognize the activities over a long video sequence becomes quite attractive. Here this is formulated as a novel N-class, model based shot detection problem which uses a stochastic, asymptotically optimal procedure, so that neither changes in content nor the types of shot transition hinder the decision making process.

We employ the HMM framework used in section 4.3, as such, for this purpose also. However this method is very different in approach and always yields scene cuts with the minimum delay performance. Here we do not attempt to train the model with the types of transition expected. There are many types of scene transitions found, which have already been dealt with in Section 4.1 of this chapter. Our objective is to detect the scene change with minimum delay irrespective of the type of transition. Although we use HMM for shot detection, the discussed method learns the activities in a given scene and makes no attempt to learn the statistics during transition, making it equally amenable to deal with all types of shot changes. The key concept is that whenever the activity in a scene changes, the data statistics changes and we capture the change point optimally (with respect to detection delay) using the Shiryaev-Robert statistical test [132]. Since the method is based on detecting changes in activities in the scene rather than intensity variation across consecutive frames, it can handle different types of shot changes like wipe, dissolve, etc. However it may be noted that the activities in the scene must be learnt before application.

4.4.1 Optimal transition detection

Hidden Markov model can be used to identify isolated activities as discussed in section 4.3. The test sequences that are fed into the model are required to be of a single temporal segment. If the sequence consists of more than a single scene, one followed by another, the recognition process will not give the correct result. Hence, we need some method to detect these scene changes automatically and continuously with the minimum possible delay and then segment the test sequence at the transition point. Thus, these temporally segmented sequences can then be fed into HMM individually so as to recognize that particular scene.

Consider the interesting formulation of the change point detection problem. There is a sequence of observations whose distribution changes at some unknown point in time and the goal is to detect this change as soon as possible, subject to certain false alarm constraints [132]. Let the unknown point of time at which the change occurs be ζ. At ζ, all or most of the components of the observation vector change their distribution. Let us consider that there are only two classes, c_0 and c_1 and the change takes place from class c_0 to c_1. If the change point occurs at $\zeta = k$, then the jth component of the feature vector $x_{j,1}, \ldots, x_{j,k-1}$ follow the distribution whose conditional density is $f_{c_0,i}^{(j)}(x_{j,i}|x_{j,1}, \ldots, x_{j,i-1})$, $i = 1, \ldots, k-1$, while the data $x_{j,k}, x_{j,k+1}, \ldots$ have the conditional densities $f_{c_1,i}^{(j)}(x_{j,i}|x_{j,1}, \ldots, x_{j,i-1})$, $\forall \ i \geq k$. We use the same feature set as in Section 4.3.

Veeravalli [147] proposed the centralized sequential change point detection procedure with a stopping time τ for an observed sequence $\{\mathbf{X}^n\}_{n \geq 1}$, where $\mathbf{X}^n = \{X_1^n, \ldots, X_I^n\}$ and $X_i^n = \{x_{j,1}, \ldots, x_{j,n}\}$. Thus $\mathbf{X}^n$ is the accumulated history of all features up to the given time instant. A false alarm is raised whenever the detection is declared before the change occurs, i.e. when $\tau < \zeta$, where τ is the computed detection time. A good change point detection procedure should give stochastically, a small detection delay $(\tau - \zeta)$ provided there are no or very few false alarms. The change point ζ is assumed to be a random variable with some prior probability distribution $p_k = P(\zeta = k)$, $k = 1, 2, \ldots$. It follows from the work of [133] that if the distribution of the change point is geometric, then the optimal detection procedure is the one that raises an alarm at the first time such that the posterior probability p_n of occurrence of change point exceeds some threshold θ, where the threshold θ is chosen in such a way that probability of false alarm (PFA) does not exceed a predefined value α, $0 \leq \alpha \leq 1$. Thus the optimal change point detection procedure is described by the stopping time

$$v = \inf_n \{n \geq 1 : p_n \geq \theta\} \tag{4.4}$$

subject to

$$PFA(v(\theta)) \leq 1 - \theta, \ 0 < \theta < 1 \tag{4.5}$$

The exact match of the false alarm probability is related to the estimation of the overshoot in the stopping rule. Putting $\theta \leq 1 - \alpha$ gives an optimal solution to this problem. Now, let us assume that the prior distribution of the change point is geometric with the parameter ρ, $0 < \rho < 1$, i.e.

$$p_k = P(\zeta = k) = \rho(1 - \rho)^{(k-1)} \ for \ k = 1, 2, \ldots \tag{4.6}$$

Shiryaev [132] defined the following two statistics for $k \leq n$. Let

$$\Lambda_n^k = \prod_{t=k}^{n} \prod_{j=1}^{J} \frac{f_{c_1,t}^{(j)}(x_{j,t}|X_j^{t-1})}{f_{c_0,t}^{(j)}(x_{i,t}|X_j^{t-1})}. \tag{4.7}$$

where the term on the right hand side can be interpreted as likelihood ratio and

$$R_{\rho,n} = \sum_{k=1}^{n} (1 - \rho)^{(k-1-n)} \Lambda_n^k. \tag{4.8}$$

Taking into account that $R_{\rho,n} = p_n[(1 - p_n)\rho]$, the Shiryaev stopping rule can be written in the following form

$$v_{\theta'} = \inf\{n \geq 1 : R_{\rho,n} \geq \theta'\}, \ \theta' = \frac{\theta}{(1 - \theta)\rho}. \tag{4.9}$$

where $v_{\theta'}$ denotes the time instant of shot change for a specific value of θ'. Shiryaev procedure is optimal in the iid case. However, [147] showed that it is asymptotically optimal when α approaches to zero under fairly general conditions.

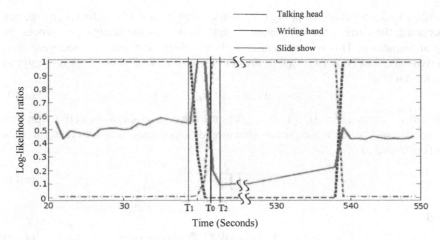

Fig. 4.3 Likelihood curves of the instructional scenes at change point. Each curve represents the likelihood normalized with respect to that of the current activity. The symbol $\int\int$ represent broken time axis.

4.4.2 The overall algorithm

In the above sections we described various components that we used in our work. In this section we show how we combined each of these components so as to use it in our framework. We also show how our framework can be used for detecting the change points with the minimum delay. Before the change occurs, the output of the proposed algorithm would be same as that of an HMM based recognizer. Hence, we can safely assume that we know the scene before the shot change occurs. Once the change occurs, our goal is to accurately detect the change point between the scenes with the minimum possible delay and very low false positives.

As mentioned earlier [147] used the Shiryaev-Robert change point detection procedure [110] to solve a two class problem for a distributed sensor network. Unfortunately, the problem of recognizing scene changes cannot be modeled as a two-class problem as there could be several types of shots representing different activities. Hence we modify their algorithm so that it can find transitions among N different classes, thereby making it an N-class change point detection procedure. For each point in time, we calculate the likelihood ratios for every category, the denominator being the likelihood of the present scene. Hence, the highest value among all these ratios is one at any given time instant and it corresponds to that of the present activity. All other likelihood ratios are less than one. The scenario remains the same until a change point occurs. When the scene changes, the likelihood curve of the previous activity drops down and the likelihood of another activity rises. This is illustrated in Fig. 4.3 where the likelihood ratios for transition shown in Fig. 4.2(b) are plotted as a function of time. In the curve, change starts at frame T_1 and the transition phase gets over at frame T_2 (ground truthed manually). The detected transition frame T_0 should satisfy $T_1 < T_0 < T_2$. We declare this point T_0 as a valid change point according to the Shiryaev stopping rule. Once it has been declared that a change has occurred, the current frame is taken as first frame and the recognition process is again initialized. This is done continuously to achieve a continuous recognition of activity and shot boundary. Hence, keeping all things unchanged, equation (4.9) can be modified to

$$v_B = \inf\{n \geq 1 : R^{c_i}_{\rho,n} \geq \theta'\}. \tag{4.10}$$

for any c_i, where $c_i\, i = \{i = 1, 2, \ldots, N\}$s are the categories of the activity present in the dataset. Let c_0 be the present shot already recognized. The Shiryaev statistics gets modified to

$$\Lambda^k_n(c_i; c_0) = \prod_{t=k}^{n} \prod_{j=1}^{J} \frac{f^{(j)}_{c_i,t}(x_{j,t}|X^{t-1}_j)}{f^{(j)}_{c_0,t}(x_{j,t}|X^{t-1}_j)}. \tag{4.11}$$

and

$$R^{c_i}_{\rho,n} = \sum_{k=1}^{n} (1-\rho)^{(k-1-n)} \Lambda^k_n(c_i; c_0). \tag{4.12}$$

Referring to Fig. 4.3 again, we explain the formulation of our algorithm. The figure shows an example of successful detection of change point. The x-axis being the time and the y-axis is the likelihood ratio. The colored dashed curves show the likelihood ratios of different activities. The vertical lines shows the actual transition boundaries. It is denoted by points T_1 and T_2 along the time line. We can see from the plot that initially when the sequence consisted of only one activity, the blue colored dashed curve, corresponding to the activity at that instant has the highest value. At change point, the likelihood for that particular activity decreases and the likelihood for the new activity increases. Hence, the likelihood ratio for the new activity increases. As this increase in ratio crosses the threshold, we initialize the value of the first frame to current frame and again the log-likelihoods are calculated using the HMM discussed in section 4.3. In Fig. 4.3 we reinitialize the likelihoods at point T_0. The new activity is shown by the rise in the value of another curve to unity.

Having arrived at the N-class transition detection solution, we now provide a step-by-step description of the change detection algorithm:

Step 1:   Compute the feature vectors for a block of test frames and feed it to the trained HMM models to recognize an activity.
         For each subsequent frame, do:
Step 2:   Compute the Shiryaev's statistics using likelihood ratios to check whether the ratio exceeds the threshold.
Step 3:   If *not* go to Step 2; else reinitialize current frame as first frame and go to Step 1.

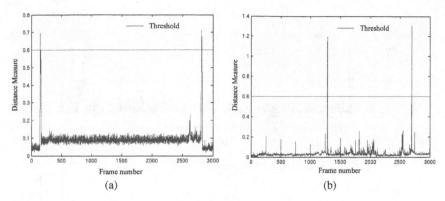

Fig. 4.4 Plot of the histogram difference measure for a lecture video containing (a) both handwritten and electronic slides, (b) only handwritten slides. The horizontal line indicates the threshold for shot detection.

4.5 Illustrative results

We have experimented extensively on six lecture videos of different instructors. The frame rate for the video is 25 frames per second (fps). The content portions in some of the videos comprise of both printed and hand written slides. Some of the videos consist only of hand written pages while the rest contains only slide shows (electronic slides).

We start with the results of shot detection and recognition which are discussed in Section 4.2 and Section 4.3. Temporal segmentation has effectively identified the scene breaks in the lecture video sequence. The plots of the histogram difference measure, obtained from two videos are shown in Fig. 4.4 (a) and (b). The spikes in these plots represent the possible scene breaks. These are detected against a threshold to effect the segmentation. As it can be seen, this histogram difference measure works well, as there is very little camera and object movement. In the case of activity recognition, features are computed as explained in Section 4.3. These features are plotted in figure 4.5 for a video segment which has two transitions. Figure 4.5(a) and (b) show the features $x_1(t)$, $x_2(t)$ respectively. Note that these features show considerable variation in characteristics during the different classes of activities which directly indicate their relevance. The training of HMM was done with the three classes of video activities as already mentioned of which four sample frames are given in Fig. 4.6. Then this trained HMM is employed for testing during which it could effectively classify a given test sequence. This is achieved by calculating the likelihood of each class when the test data from the video frame sequence is given to the trained HMMs.

Now we show the results obtained using the method of optimal shot detection and recognition using Shiryaev-Robert statistics. The likelihood ratios of the three classes of instructional activity are the key factors of decision making, of which a typical diagram is already given in Fig 4.3. Referring to it again, it can be seen that

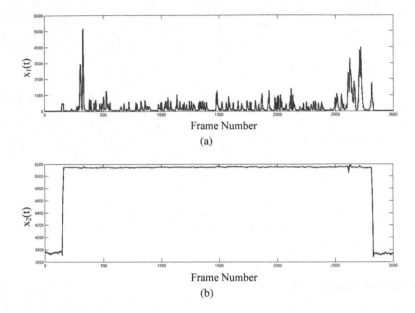

Fig. 4.5 Plots of the features used for shot detection in a video segment. (a) and (b) are the features $x_1(t)$ and $x_2(t)$, respectively. Note the significant difference in their characteristics before and after transitions.

when a scene change occurs, the likelihood curve of one activity drops down and that of another activity rises. For illustration, the log-likelihood ratios for the three classes for the temporal duration as shown in figure 4.5 are plotted in figure 4.7(a). The green curve denotes talking head, the red one denotes slide show and the blue one denotes writing hand. Note that at frame number 152, the scene transits from slide show to talking head and at frame number 2823, it transits back to slide show. For a comparative study, the simple histogram difference for the same temporal duration is plotted in figure 4.7(b), in which it crosses the threshold a couple of times during both the transitions, signifying certain temporal ambiguities as regards when exactly the transition takes place. This is because both the transitions are dissolves with a duration of 10 - 15 frames. To make it clear, the first transition in the given temporal duration is shown in a larger time scale in figure 4.8. The histogram difference curve in figure 4.8(a) shows considerably high values during the dissolve transition and so there is an ambiguity in change detection. But the log-likelihood curves shown in figure 4.8(b) do not suffer from such problems and are able to fix the change point without any ambiguity. Hence the change detection based method performs quite well irrespective of the type of transition involved. This is because the algorithm automatically recognizes the activity in a video and continuously looks for a change point thereafter.

Now we demonstrate the performance of this method in terms of false alarm and detection delay. The experiments were aimed at finding how well the algorithm per-

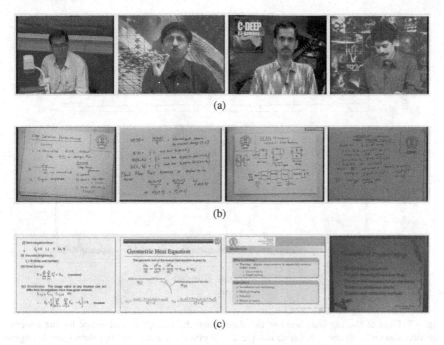

Fig. 4.6 Sample frames selected from four sequences used for training HMM. (a) for talking head sequence, (b) for writing hand sequence, and (c) for slide show sequence

forms with respect to false alarm and detection lag when the probability of false alarm is changed. The dataset consisted of forty two scene transitions having all possible combinations and different types of shot changes. The duration of the transition varied from a minimum of two frames to a maximum of seventeen frames. The ground truths were obtained through subjective reviewing. The transition detection algorithm takes the likelihoods computed by HMMs in each time frame and computes the probability of transition in linear time.

In the testing phase, the experiments were carried out for different values of threshold α. When α was kept as low as about 0.02, the results showed a large number of false alarms. We increased α in steps of 0.02. The accuracy increases as α is increased. For $\alpha = 0.07$ only two out of forty two showed false alarms. The remaining sequences gave correct results with no false alarm. When the value of the threshold was further increased to 0.1, all the sequences were detected perfectly and no false alarm was seen in any of the sequences. The average number of false alarms for the experimented scenes with respect to different values of α is plotted in Fig. 4.9. The gradually decreasing plot intuitively substantiates the claim of proper working of the change point detection algorithm.

On the similar lines of the above experimentations, we tested for the variations in detection delay given by $(T_0 - T_1)$ against different values of threshold as shown in figure 4.9. With lower values of α though there were frequent false alarms, the de-

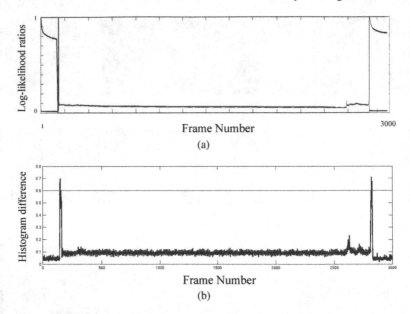

Fig. 4.7 Plots of the log-likelihood of the classes and the histogram difference for the temporal duration shown in figure 4.5. (a) shows the log-likelihood curves and (b) shows the histogram difference.

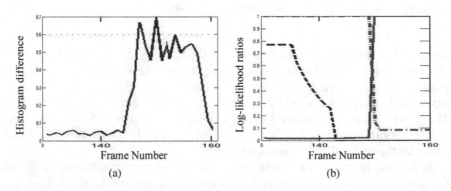

Fig. 4.8 Plots of the histogram difference and the log-likelihood ratios of the classes in a larger time scale for the first transition as given in figure 4.7. (a) shows the histogram difference and (b) shows the log-likelihood curves.

tection delays were considerably small. With the increase in α, the detection delay also increases. When α was initialized to 0.02, it gave very low values of detection delays. There is a marginal increase in the detection lag when the threshold is increased to 0.05. On further increase to $\alpha = 0.1$, the average delay increases to almost twice of the detection lags at $\alpha = 0.05$. Hence, if we keep the value of α low, we get

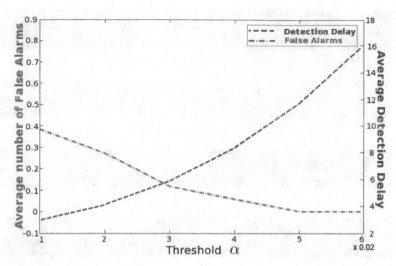

Fig. 4.9 Plots showing average number of false alarms and detection delay for different values of threshold α.

very accurate and a low detection delay, but it increases the occurrence of the false alarms. When α is kept high, detection accuracy improves, but the lag increases.

Finally in Fig. 4.10 we give a few illustrative results on change detection on different sequences at shot boundaries. Here only one example thumbnail of boundary frame sequence for each video is shown and the corresponding detected scene change is denoted by a large gap between thumbnails. Note that in Fig. 4.10(a) the scene transition is a cut while in (b) - (f) it is dissolve which occur through eight frames in (b) - (e) and through twelve frames in (f). Also, figures 4.10(a) - (d) show transitions between talking head and writing hand while (e) - (f) show those between talking head and slide show. Regardless of the type of scene change, the change detection algorithm effectively identified shot breaks at the points marked, differentiating the two activities on its two sides.

4.6 Discussions

This chapter introduced the preliminary procedures concerned with the higher level analysis of instructional videos. Two closely related tasks, namely shot detection and shot recognition are discussed. We started with a method of shot detection based on the histogram difference in intensity space. Then an HMM based method was suggested for the classification of activities in-between the shot boundaries. In the joint shot detection and recognition method, the HMM framework designed for the shot classification itself performs the change detection, through a continuous monitoring of the likelihood functions. This framework is capable of automatically detecting

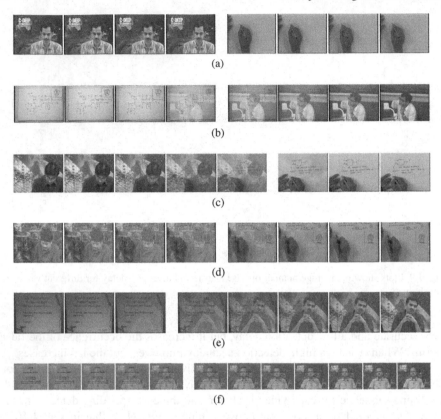

Fig. 4.10 Example results showing thumbnail frame sequences at shot boundaries for different videos and the obtained change points. A large gap between thumbnails shows the scene break.

the transitions in scenes, thereby separating them at the transition points so that the individual activities can be efficiently and continuously recognized with a guaranteed minimum delay. Using these methods the instructional activities are classified into three categories: (1) talking head, (2) writing hand and (3) slide show. We give a nomenclature of *content frames* to the second and the third classes as a whole and *non-content frames* to the class of talking head frames. Further analysis on these classes are to be performed according to the characteristics of the constituent frames as described in detail in following chapters.

It has been observed that the histogram difference is a good measure for temporal segmentation since both the global and object motion in instructional scenes are very low. So a simple and fast scene change detection is performed by using this measure. Also it can be seen that the HMM classifier effectively categorized the instructional activities, using only two relevant features as the scene complexity is quite low for lecture videos. Even though the transition examples used here are taken from lecture video, the suggested model for optimal shot detection and recognition works equally well for other videos, but what makes it different is the choice of features.

The feature selection process differs for various classes of videos like educational video, news video, sports video, etc. Hence the results are quite applicable to all types of video irrespective of its scene composition.

This preprocessing is very essential for almost all higher level analysis of videos since the information about scenes renders a good platform for the design of efficient algorithms which target the classes separately. The next chapter will discuss a method of repackaging of lecture videos and its content re-creation by analyzing the already classified video segments.

Chapter 5
Video Content Re-creation

5.1 Introduction

The wide spread of digital video and the increased use of Internet by the student community have brought many new applications in the field of education. In this scenario, research and development in new multimedia technologies which focus on the improved accessibility of the huge volume of the educational video data have become quite crucial. The related works mainly make use of the fact that digital video contains redundant and unstructured data streams that span along the time sequence. In view of fast browsing, limited bandwidth and reduced storage, recent research has led to many approaches for understanding and summarizing commercial video [42, 74]. However, a very little has been achieved till date that can efficiently encode hours of educational video into only a few megabytes of data, thus enhancing the outreach of the distance education programs. Our effort in this monograph lies in this direction by building a completely automated system without much sacrificing the pedagogy. The production, storage and accessing of lecture video can undergo revolutionary changes if one could remove the temporal and spatial redundancies associated with pedagogical activities and present it in a compact and semantically correct way, without any loss in pedagogic content or the effectiveness of the lecture video. It is the motivation of this monograph. The key idea is that a programmed display of the representative key-frames from the different segments of the video along with the synchronous playback of audio will render a repackaged representation of the original video, without much loss in instructional values.

A task closely related to the repackaging problem discussed above is known as *media retargeting* [62, 73, 130]. It aims at preserving the viewers' perceptual experience by suitably maintaining the importance of information contained in the media. Most often, this implies an appropriate display of a given image/video in the changed display size. Another area related to video transmission and streaming is *transcoding* [1, 161, 166] wherein the display resolution is changed based on the type of device, to which the media is streamed. However, transcoding is not relevant to this work since resolution conversion is not the key issue in solving our problem. Most of the retargeting methods found in literature for conventional videos [62, 73, 130] use cropping, scaling, seam-carving, and warping, which drop pixels or frames in some manner to effect the desired adaptation. Since the object motion in instructional scenes are much less, these methods do not hold good for lecture video. Also, the legibility of the text portions on the content frames are to be preserved, without any cropping and scaling down. Hence we adopt a different strategy such that a key-frame which is defined as a completely written slide is retained for display throughout the temporal location where all the previous redundant frames belonging to that content key frame are removed. In addition, we propose to use a super-resolved content key-frame for improved legibility. A key-frame corresponding to the talking head sequence is displayed as such for the required temporal segment. In a nutshell, our method comprises of two tasks, the first one being the analysis part by which we summarize the lecture video into key-frames, audio and related meta-data and the second one being the media re-creation to display these key-frames along with the audio according to the meta-data. Since the video frames corresponding to talking head and written/electronic slide categories differ much in semantics from the talking head sequence, these classes are to be processed with appropriate key-frame selection mechanisms separately, for effective representation of the original video.

To date, several efforts have been attempted to summarize commercial videos. Note that the concept of video summarization as propounded in [114] is not at all useful in dealing with instructional video as their method yields physically incorrect (in space and time) results. Since segmentation based video summarization methods [171] depend on syntactic structures, they are not applicable to the instructional videos. Due to the computational overhead, clustering based method [175] and set-cover method [20] are also not suitable in this scenario together with the fact that the frame distance definition based on histograms does not show a content difference in instructional videos. Hence a content based key-frame selection method [85] is to be used for instructional videos, by which it can address the issues concerned with the activities like writing hand or electronic slide presentations. The number of *ink pixels* (pixels corresponding to scribbling on paper or blackboard) in a frame can be used as a measure of its video content, which may be employed for the key frame selection by maximizing the content coverage and minimizing motion and content occlusion.

The method proposed in [84] treats ink pixels as the semantic content of an instructional video frame which is used for the key-frame selection. Since it directly process ink pixels in spatial co-ordinates, it does not take the advantage of the fact that the written regions appear almost horizontal in document images. Consequently, a method which uses the projection of ink pixels horizontally would provide a simpler and faster method of key-frame selection. The novelty of our work lies in the idea of applying the horizontal projection profile (HPP) of the ink pixels for key-frame extraction of content segments in the instructional video, which is a simple and efficient one, compared to existing methods. In case of non-content segments like talking head, a perceptual visual quality alone should suffice to provide a sense

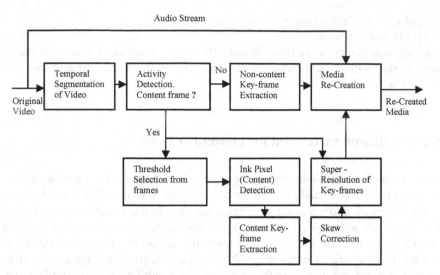

Fig. 5.1 Block schematic of the designed system for pedagogic repackaging of instructional video.

of look and feel of the instructor to the viewers. Hence a no-reference visual quality assessment scheme [154] is used for the key-frame selection of non-content frames.

As the instructor writes on a piece of paper for illustration, the paper undergoes occasional inadvertent skews. A skew correction mechanism for content key-frames based on Radon transform is suggested. The resulting distinct content key-frames are super-resolved based on the motion information before we re-create the instructional media. This provides better legibility of the written texts. These key-frames, together with some of the talking head key frames and the associated audio, yield a repackaged lecture video sequence called instructional media package (IMP). Since the temporally long-lasting video shots have been replaced by the corresponding static key frames, the repackaged media offers a large reduction in the data size. During media re-creation we display these key frames at proper temporal locations along with the audio playback in complete synchronization.

We started this chapter by placing the work in the context of existing literature. Now we describe the suggested method of pedagogic repackaging and content re-creation of instructional video of which the details will appear in the following sections. Finally we will show detailed experimental results and its analysis in section 5.7 and will summarize in section 5.8.

Our objective is to repackage an instructional video and to re-create its pedagogic content without much loss in instructional values. For this, we have built a complete system, the block schematic of which is given in Fig. 5.1. It works with the following distinct steps for the repackaging of the lecture video:

(a) Shot detection and recognition
(b) Key-frame extraction for content frames
(c) Redundancy elimination in content key-frames
(d) Content super-resolution

(e) Key-frame extraction for non-content frames, and

(f) Media re-creation.

Since the shot detection and recognition methods are already dealt with in Chapter 4, here we start with the key-frame selection for content frames i.e., writing hand and slide show frames.

5.2 Key-frame extraction for content frames

The key frame extraction in content segments of the lecture video has two phases - the ink pixel detection and the HPP based key-frame selection. Since the discriminating information to select the key-frames lies only in the semantics of the written text and hence in its spatial location in the frame, the slow variation of intensity in the background can be conveniently ignored. This is the philosophy behind employing ink pixel detection before key-frame extraction in content frames. To reduce computation, we select only one frame in every 2 seconds i.e., we sub-sample frames by a factor of 50 as the original video runs at 25 fps. This is under the assumption that there is not much variation in visual content during a 2 second duration, as is common in most instructional video.

5.2.1 Ink-pixel detection in content frames

Since the key-frame identification or pedagogic summarization step is to be applied to the content classes also, some measure of its semantic content should be defined. This is achieved through the preliminary step of ink detection, which is done using the well-known histogram thresholding technique [65, 99]. By this, the pixels in the content frames are converted into *ink* and *paper* pixels, corresponding to 0 and 255 values of gray levels, respectively, according to the equation

$$B(m,n) = \begin{cases} 0 & if\ I(m,n) < T \\ 255 & otherwise. \end{cases} \qquad (5.1)$$

Here a suitable threshold T is needed for separating the foreground and background in the content frames. For this, we start with the gray-level histogram $h(i)$ of the content frame which is modeled by a bimodal Gaussian mixture model (GMM). Its probability density function (PDF) is given by

$$p(i) = \frac{\varepsilon}{\sqrt{2\pi}\,\sigma_1} e^{-\frac{1}{2}\left(\frac{i-\mu_1}{\sigma_1}\right)^2} + \frac{1-\varepsilon}{\sqrt{2\pi}\,\sigma_2} e^{-\frac{1}{2}\left(\frac{i-\mu_2}{\sigma_2}\right)^2} \qquad (5.2)$$

where i is the intensity level, ε is the proportion of the mixture, μ_1 is the foreground mean, μ_2 is the background mean, σ_1^2 is the foreground variance and σ_2^2 is the back-

ground variance. We employ the Kullback Leibler (KL) divergence [64] for computing the threshold. We minimize the KL divergence J from the observed histogram
$h(i)$ to the unknown mixture distribution, $p(i)$ to estimate the model parameters. J
is given by

$$J = \sum_{i=0}^{255} h(i) \log \left[\frac{h(i)}{p(i)} \right]. \tag{5.3}$$

Expanding the above equation we get

$$J = \sum_{i=0}^{255} h(i) \log[h(i)] - \sum_{i=0}^{255} h(i) \log[p(i)]. \tag{5.4}$$

Since the first term does not depend on the unknown parameters, the minimization
is equivalent to minimizing the second term. We denote this term as the information
measure Q. Hence

$$Q = -\sum_{i=0}^{255} h(i) \log[p(i)]. \tag{5.5}$$

To carry out the minimization, we assume that the modes are well separated. If T is
the threshold which separates the two modes, we have

$$p(i) \approx \begin{cases} \frac{\varepsilon}{\sqrt{2\pi}\sigma_1} e^{-\frac{1}{2}\left(\frac{i-\mu_1}{\sigma_1}\right)^2}; & 0 \leq i \leq T \\ \frac{1-\varepsilon}{\sqrt{2\pi}\sigma_2} e^{-\frac{1}{2}\left(\frac{i-\mu_2}{\sigma_2}\right)^2}; & T < i \leq 255 \end{cases}$$

Now

$$Q(T) = -\sum_{i=0}^{T} h(i) \log \left[\frac{\varepsilon}{\sqrt{2\pi}\sigma_1} e^{-\frac{1}{2}\left(\frac{i-\mu_1}{\sigma_1}\right)^2} \right]$$
$$- \sum_{i=T+1}^{255} h(i) \log \left[\frac{1-\varepsilon}{\sqrt{2\pi}\sigma_2} e^{-\frac{1}{2}\left(\frac{i-\mu_2}{\sigma_2}\right)^2} \right]. \tag{5.6}$$

The assumption of well separated modes means that the mean and variance esti-
mated from $h(i)$ for $0 \leq i \leq T$ will be close to the true mean and variance μ_1 and σ_1^2.
Similarly, the mean and variance estimated from $h(i)$ for $T < i \leq 255$ will be close to
the true mean and variance μ_2 and σ_2^2. The value $T = T^*$ that minimizes $Q(T)$ is the
best threshold. Hence we minimize $Q(T)$ with respect to $\{\mu_1, \sigma_1, \mu_2, \sigma_2, \varepsilon, T\}$ and
compute the optimum threshold required for ink pixel detection of content frames.
The ink pixel detected frame $B_t(m,n)$, obtained by using equation (5.1) is then employed for the HPP based key-frame selection.

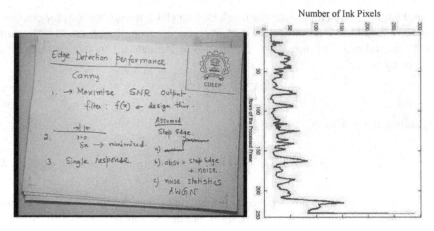

Fig. 5.2 A content frame and its HPP. Note the horizontal excursions in the HPP which result from well written horizontal lines in the content frame.

5.2.2 HPP based key-frame selection

A content segment is usually a long duration one but of low visual activity shot in which a particular academic topic is explained to the students. Hence some quantitative measures are needed to select an appropriate number of key-frames in the video segment. We employ HPP for this purpose. HPP is a well tested concept in document analysis [6, 70, 137]. It is defined as a vector, with each element obtained by summing the pixel values along a row. We compute the HPP for each ink pixel detected frame by projecting the ink pixels in a frame on the y-axis as shown in Fig. 5.2. Since the frame size is $M \times N$, its HPP is an array F_t of length M in which each element stands for the total number of ink pixels in a row of the processed frame. This is obtained by the equation

$$F_t(m) = \frac{1}{255} \sum_{n=1}^{N} |B_t(m,n) - 255|; \quad m = 1, 2, ..M. \tag{5.7}$$

We suggest the use of HPP based key-frame extraction for content frames, since text regions in the image frame appear almost horizontally. But this is under the assumption that the document does not suffer from any skew, which is relaxed in the next section. From the instructor's perspective, his/her writing should be nearly horizontal and there should be no movement of the page as the writing proceeds. Depending upon the teaching habits of the instructor, there may be two cases:

(i) Same slide (handwritten or electronic) is being used and no additional material is being written or shown on it

(ii) Some additional material is being written or displayed on the same paper or slide, so what matters is the picture of this at the end of writing on the same page.

In the first case, the content of the video frame does not vary, but in the second one, definitely, there is an increase in content.

For similar handwritten slides, printed or electronic slides, the projection profiles for consecutive frames differ only negligibly. Hence variation in the projection profile is a direct measure of dissimilarity in content. It is used for the identification of key-frame in our algorithm. We define a parameter S_1 for this purpose of quantifying local dissimilarity in the HPP

$$S_1 = \sum_{m=1}^{M_1} | F_t(m) - F_{t-1}(m) | \qquad (5.8)$$

where $M_1 \leq M$ is the extent (rows) up to which the projection profiles of consecutive frames are assumed to overlap. If S_1 is greater than an empirically chosen threshold (T_{S_1}), the content of the current frame is said to have deviated much from that of the previous one, which indicates the introduction of a new slide or page. At this moment, the last frame is declared as the key-frame. Otherwise, the consecutive frames should belong to the same writing hand page sequence or slide show (electronic slides). Whenever the writing hand enters a new page or electronic slide transits to a new one, it is detected. Note that the difference is taken not for the entire set of rows, but only up to those rows which have been filled up by written materials progressively. This is easily detectable from the HPP. This process is required to effectively check the similarity in the beginning lines of a page or slide.

The total number of ink pixels in a frame (S_2) is treated as a measure of pedagogic content of the frame, which can be used as a measure to verify how useful these content frames are.

$$S_2 = \sum_{m=1}^{M} F_t(m). \qquad (5.9)$$

That is, only if the content exceeds some pre-determined threshold (T_{S_2}), one requires the key-frame selection. This threshold value is calculated from the distribution of the total number of ink pixels among the processed frames. This eliminates the selection of a frame resulting from a blank page at the beginning of writing and a moving hand of the instructor occasionally obscuring it. Similarly, any occlusion by a writing hand at the bottom of hand-written slides can be effectively eliminated by introducing a third parameter, S_3 and checking against a threshold (T_{S_3}) which is obtained from the distribution of HPPs of the processed frames.

$$S_3 = \sum_{m=M_1}^{M} | F_t(m) - F_{t-1}(m) | . \qquad (5.10)$$

Comparing equation (5.8) with equation (5.10), we observe that S_1 gives the measure of dissimilarity in the similar part of the frame and S_3 measures the dissimilarity in the dissimilar part of the content frame. Hand-written slides, totally free from occlusion from hand or other body parts or its shadows, are known as *clean frames*. This algorithm searches for clean frames, if any, otherwise it yields those frames, with the minimum occlusion. A frame is chosen as the corresponding clean key frame if it satisfies

$$(S_1 \geq T_{S_1}) \& (S_2 \geq T_{S_2}) \& (S_3 \leq T_{S_3}).$$

The last condition $S_3 \leq T_{S_3}$ is not required for the key-frame selection of electronic slide presentation as the corresponding slides are fed directly from the computer to the screen and there is no scope for occlusion of contents.

5.3 Redundancy elimination in content key-frames

Multiple key-frames depicting the scene corresponding to the same textual contents can happen if there exists a skew for the content frame, resulting from any tilt of the writing page by the instructor or by unwanted vertical movement of the paper during writing. This issue is less troublesome in case of slide shows. Those key-frames obtained by the method explained in the previous sub-section are called candidate key-frames, as they contain duplicate key-frames resulting from the translation or rotation of the writing page.

The problem of having redundancies in key frames are effectively overcome by incorporating a skew correction mechanism in our algorithm. This is done by computing the Radon transforms (RT) [33, 61] of the candidate key frames. The RT of a frame I(x,y) is given by

$$g(s, \phi) = \int_{-\infty}^{+\infty} \int_{-\infty}^{+\infty} I(x, y) \delta(x \cos \phi + y \sin \phi - s) dx dy \qquad (5.11)$$

$$-\infty < s < +\infty, 0 \leq \phi < \pi$$

where s is the distance variable and ϕ is the direction of projection as shown in Fig. 5.3. Hence $g(s, \phi)$ is the 1-D projection of the image at angle ϕ. It gives the *ray-sum* of the image at an angle ϕ for varying s. Each point in the (s, ϕ) space corresponds to a line in the spatial domain (x, y). Note that HPP defined in equation (5.7) is same as $g(s, \phi)$ for $\phi = 0$ for the ink pixel detected frame. If two video frames represent the same visual content, but with different skew, their RT along some value of ϕ will have high correlation. This principle is used for the elimination of the duplicate key-frames of the content frames.

A reader may notice that we first detect the candidate key frames and then perform the skew analysis as opposed to skew correction first, which is more common in document image processing. This is due to the fact that key frame detection is a much faster operation compared to computing the Radon transform. Hence the computation is restricted to only the detected candidate frames. For the candidate key-frames, RT is computed for the range of angles, -10 to +10 degrees, in the intervals of 0.5 degrees. Correlation of RT is used for finding out the degree of similarity between the candidate key-frames. We find the correlation coefficient of the Radon

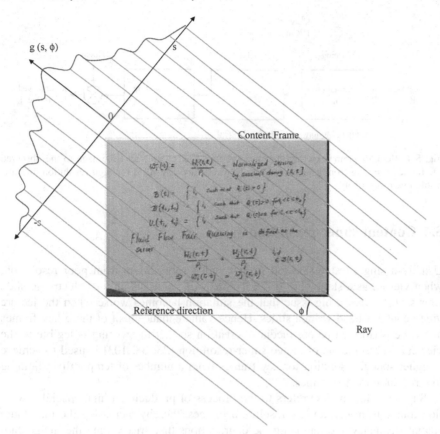

Fig. 5.3 Illustration of computation of Radon transform $g(s, \phi)$ of a candidate key frame along a direction ϕ.

transform of a candidate key-frame along a reference direction with the Radon transform of the next candidate key-frame along all 41 directions mentioned above. Let g_0 be the Radon transform of a candidate key-frame in the reference direction and $g_1, g_2, g_3, \ldots g_{41}$ be the Radon transforms of the next candidate key-frame along the directions of interest. Then the correlation coefficient is computed for all the 41 cases and if the correlation coefficient has a significant value along some particular direction, the candidate frames are not distinct. If the RT of a candidate key-frame does not have any similarity with those of the previous ones for $| \phi | < 10^o$, it is declared as a distinct key-frame else the candidate key frame is discarded. This approach of the computation of RT in a limited angular span can still yield good results as there is little chance for a lecture video frame to have a skew more than ± 10 degrees.

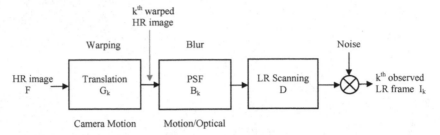

Fig. 5.4 Block diagram depicting the observation model relating an HR image to the observed LR frames resulting from a video camera. Here HR and LR stand for high resolution and low resolution, respectively.

5.4 Content super-resolution

Usually a single frame selected from a video shot suffers from poor resolution, when viewed as a static one. If the lecture video consists only of electronic slide shows, this is not a big issue, but the situation becomes worse when the lecture video contains hand-written slides. Hence some enhancement of these key-frames are to be performed before media re-creation so that the writing is legible to the viewers. The principle of image super-resolution [23, 38, 103] is used to achieve a higher spatial resolution for key-frames from a number of temporally adjoining low-resolution video frames.

Super-resolution (SR) refers to the process of producing a high spacial resolution image from several low resolution images, thereby increasing the maximum spatial frequency and removing the degradations that arise during the image capturing process using a low resolution camera. In order to obtain super-resolution we must look for non-redundant information among the various frames in an image sequence. The most obvious method for this seems to be to capture multiple low resolution observations of the same scene through subpixel shifts due to the camera motion. If the low resolution image shifts are in integer units, then there is no additional information available from subsequent low resolution observations for super-resolution process. However, if they have subpixel shifts then each low resolution, aliased frame contains additional information that can be used for high resolution reconstruction. The success of any super-resolution reconstruction method is based on the correctness of the low resolution image formation model that relates the original high resolution image to the observed images. The most common model used is based on observations which are shifted, blurred and decimated (and hence possibly aliased) versions of the high resolution image. The observation model relating a high resolution image to low resolution video frames is shown in Fig 5.4. In the figure $F(m, n)$ is the desired high resolution image which is obtained by sampling the spatially continuous scene at a rate higher than or equal to the Nyquist rate. The camera motion at the kth time instant during the exposure can be modeled as pure translation t_k for the distance education video as the paper on which the instructor

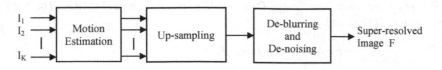

Fig. 5.5 Scheme for super-resolution from multiple subpixel shifted observations.

writes is perpendicular to the camera axis. The blurring which may be caused by the optical system or due to relative motion between camera and the scene, can be modeled as linear space invariant. One can select an appropriate point spread function (PSF) for the blur. These warped and blurred high resolution images undergo a low resolution scanning, i.e., sub-sampling or decimation, followed by noise addition, yielding the low resolution observations.

Let I_k, $k = 1, 2, ...K$ be a sequence of video frames adjoining a particular content frame. As mentioned, these images are regarded as different low resolution representations of a single high-resolution image F of the instructional scene. More specifically, each observed image I_k is the result of a geometric warping, blurring and decimation performed on F. Mathematically, the whole process is given by

$$I_k = DB_k G_k F + \eta_k; \quad for \ 1 \leq k \leq K \tag{5.12}$$

where G_k represents geometric warping performed on F due to translation t_k, B_k is the blurring matrix, D is the high resolution to low resolution decimation operator and η_k is an *i.i.d.* additive Gaussian noise with zero mean and variance σ_n^2. Unless there is a change in camera focal length or there is a variation in scene depth, the blur PSF B_k remains constant for all k. This is definitely true for the distance education video. The goal of super resolution based image enhancement and reconstruction is to recover the original image F from the observed images I_k, $k = 1, 2, ...K$.

Most of the super-resolution methods found in the literature use motion between the observed frames as a cue for estimating the high resolution image. This being the most intuitive approach for super-resolution, it is based on a three stage algorithm consisting of (a) registration or motion (G_k) estimation, (b) interpolation (inverse of D), and (c) restoration (de-blurring and de-noising) as shown in Fig 5.5. Here the low resolution observations $I_1, I_2, ...I_K$ are used as input to the registration or motion estimation module. This step is used to find the relative motion between the frames with a subpixel accuracy. The assumption here is that all the pixels from the available frames can be mapped back onto the reference frame based on the motion vector information. Next, the interpolation onto a uniform grid is done to obtain a uniformly spaced upsampled image for each of the observations. Once the upsampled images on uniformly spaced grid points are obtained, multichannel restoration is applied to remove the effects of aliasing and blurring and to reduce noise. The restoration can be performed by using any deconvolution algorithm that considers the presence of an additive noise. For the realization of the above algorithm we look for a maximum a posteriori (MAP) estimate of the high resolution image F. For the prior $p(F)$, we assume the image F to be a Markov random field (MRF). The MAP estimate [23] is given by

$$\hat{F} = \arg\max_F p(I_1, I_2 ... I_K | F) \, p(F).$$

$$= \arg\min_F \sum_k | I_k - DBG_k F |^2 / \sigma_n^2 + U(F) \qquad (5.13)$$

where $U(F)$ is an appropriate potential function defined on the image lattice. We use the smoothness criterion defined in [23] as the potential function in this book. Hence $U(f)$ is defined as

$$\lambda \sum_i \sum_j \left[(F(i,j) - F(i-1,j))^2 + (F(i,j) - F(i,j-1))^2 \right]$$

where λ is an appropriate weighting function. The low-resolution sequence is chosen around the position of the selected key-frame in the raw video, provided there is no shot change within the chosen set of frames. About 10 to 20 frames are used for super-resolving a key-frame and a magnification factor of 2 to 4, as desired, is achieved.

5.5 Key-frame extraction for non-content frames

The talking-head activity, which is visually a non-content one, is to be represented by an appropriate key-frame during media re-creation to provide a sense of situational awareness to the viewer. We need to pick up a representative frame depicting the instructor, which has the best visual quality. We employ the no-reference perceptual quality assessment method used in [154] for the visual quality assessment of the talking head sequence which in turn is used for the key-frame extraction. In this method, blurring [89] and blocking effect [153] generated during the image compression process are considered as the most significant artifacts which degrade the quality of an image. Our image frames often are in JPEG format which is a block DCT-based lossy image coding technique. It is lossy because of the quantization operation applied to the DCT coefficients in each 8×8 coding block. Both blurring and blocking artifacts are created during quantization. The blurring effect is mainly due to the loss of high frequency DCT coefficients, which smoothes the image within each block. Blocking effect occurs due to the discontinuity at block boundaries, which is generated because the quantization in JPEG is block-based and the blocks are quantized independently.

Let the frame be $I(m,n)$ for $m \in [1, M]$ and $n \in [1, N]$. A difference signal along each horizontal line is computed as

$$d_h(m,n) = | I(m, n+1) - I(m,n) |, \quad n \in [1, N-1] \qquad (5.14)$$

The blockiness is estimated as the average differences across 8×8 block boundaries

$$B_h = \frac{1}{M[(N/8)-1]} \sum_{i=1}^{M} \sum_{j=1}^{(N/8)-1} d_h(i,8j). \tag{5.15}$$

Since blurring causes a reduction in signal activity, combining the blockiness and the activity measure is more useful to deal with the relative blur in the frame. The activity is measured using two factors, the first one being the average absolute difference between in-block image samples

$$A_h = \frac{1}{7} \left(\frac{8}{M(N-1)} \sum_{i=1}^{M} \sum_{j=1}^{N-1} d_h(i,j) - B_h \right) \tag{5.16}$$

and the second one being the zero-crossing (ZC) rate. We define for $n \in [1, N-2]$

$$z_h(m,n) = \begin{cases} 1 & \text{for horizontal ZC at } d_h(m,n) \\ 0 & \text{otherwise.} \end{cases}$$

The horizontal ZC rate is estimated as

$$Z_h = \frac{1}{M(N-2)} \sum_{i=1}^{M} \sum_{j=1}^{N-2} z_h(i,j). \tag{5.17}$$

Then we calculate the corresponding vertical features B_v, A_v and Z_v by a similar procedure and finally, the combined features as

$$B = \frac{B_h + B_v}{2}, A = \frac{A_h + A_v}{2}, Z = \frac{Z_h + Z_v}{2}. \tag{5.18}$$

The quality score for a frame is given by [154]

$$Q_N = -a + bB^{-c_1}A^{c_2}Z^{c_3} \tag{5.19}$$

where a, b, c_1, c_2 and c_3 are the tuning parameters, empirically estimated using subjective test data. We use the same values as suggested in [154].

The quality score defined by equation (5.19) is computed for the frames from the talking head segment of the instructional video and the frame with the maximum quality score locally is selected as the key-frame to represent the talking head segment as a whole. The sub-sampling of frames by a factor of 50 as given in section 5.2 is employed here also to reduce computation.

5.6 Media re-creation

The key-frames in different content segments give a pedagogic representation of the summarized activity in that video paragraph of the lecture video. On media re-creation, these frames, together with a key frame in the talking head sequence

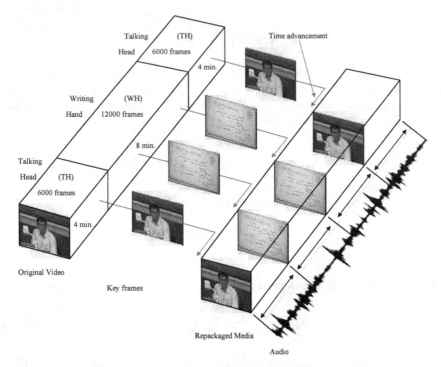

Fig. 5.6 A schematic diagram showing the sequence of the original video for 16 minutes, resulting key frames and the re-created media, along with the audio. Note the time advancement while inserting the key frames.

and the associated audio, yield a highly compressed and repackaged multimedia document representing the original lecture video. The overall compression factor may vary according to the frequency of change in electronic slides or handwritten pages. The main compression comes from the fact that video segments are replaced by high resolution static frames. A diagram showing the sequence of the original video for 16 minutes, the summarized one and the re-created one, along with the audio is given in Fig. 5.6. Here about 24000 frames are shown to be segmented into appropriate time slots: talking head (TH) and writing hand (WH) categories.

Media re-creation is done with FLASH. The talking head sequence is displayed through its representative key frame, along with the associated audio. This provides a sense of classroom awareness to the viewer, as a feel of the presence of an instructor improves the attention span of the viewer. From the previous subsections, we have noted that a key-frame which is the maximally filled content page of the writing hand or slide show sequence was detected after having completed writing the page by the instructor, while the discussion about the content of the page by the instructor had started just after completing the last working page. Hence during media re-creation, this detected key frame is placed at the beginning of the corre-

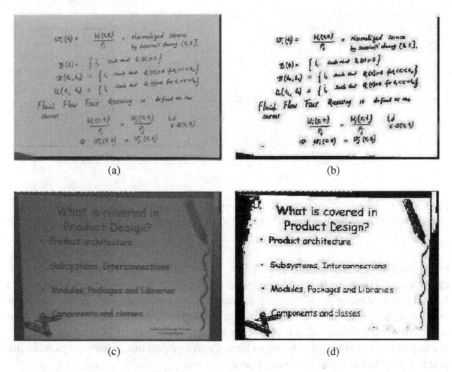

Fig. 5.7 Content frames and their corresponding ink-pixel detected frames. (b) and (d) are ink-pixel detected frames from original frames (a) and (c).

sponding time slot (called time advancement in Fig. 5.6) so that the viewer or the student can have the contents fully available at the time of audio playback.

5.7 Illustrative results

We have experimented extensively on various lecture videos by different instructors. The frame rate for the video is 25 frames per second. The content portions in some of the videos comprise of both printed and hand written slides. Some of the videos consist only of hand written pages while the rest contains only slide shows.

We first show the results regarding the key-frame extraction for content segments, i.e., writing hand sequence and slide show sequence. Here the extraction phase starts with ink pixel detection of which some sample results are shown in Fig. 5.7. Note that the possible variations of intensity in the background are eliminated, thus highlighting only the written text and its spatial arrangement. This is because one is interested only in the semantic content of the frame for key-frame selection. Usually the spatial borders of content frames contain unwanted shades/black patches (see Fig. 5.7 for example), which can be cropped before processing, thus eliminat-

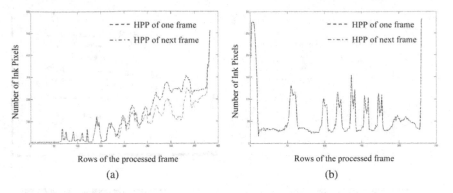

Fig. 5.8 Plot of horizontal projection profiles (HPP) of two temporally adjacent frames for a lecture video containing a (a) hand-written page and (b) slide show (electronic slide).

ing unnecessary computations. Our region of interest is only that portion, where the text appears. Hence rows or columns with about 10-20 pixel width can be removed from the boundary of the frame, before the above computations are done. The presence of such patches near the boundaries are not critical in the case of segmentation or activity recognition, since they appear almost uniformly in all frames. But during the content analysis, it is better to have them cropped out, otherwise pseudo-ink pixels get added to affect the overall performance.

The HPPs of two frames 2 second apart, plotted on the same graph for two different lecture videos are shown in Fig. 5.8. Fig. 5.8(a) gives the HPPs representing a writing hand and 5.8(b) gives those representing a slide show. On examining the HPP plots, given in Fig. 5.8(a), it can be seen that they coincide locally, through some rows, in the document as the instructor continues writing on the same sheet. If they coincide, surely, the previous frame is a subset of the next frame in terms of contents, under the assumption of no shift or skew. In Fig. 5.8(a), the HPPs deviate from each other, much towards the bottom rows indicating the continuation of a writing hand there. In Fig. 5.8(b) the HPPs almost coincide from top to bottom, since they originate from an undisturbed slide show sequence. Hence the use of HPP is found to be an effective and fast method for the key-frame extraction of lecture video.

The total number of ink pixels in a frame is treated as a crude but effective measure of the pedagogic content of that frame, the plots of which are shown in Fig. 5.9. Fig. 5.9(a) plots the content of the processed frames for a lecture video with a writing hand and 5.9(b) plots the same for a slide show. One may note the variation in pedagogic contents among the processed frames, which clearly indicates the presence of content and non-content regions in the video sequence. As can be seen, switching between content and non-content segments is fast in 5.9(b) compared to 5.9(a) as it takes a longer duration to write on the paper. A slide based lecture often tends to be fast-paced and hence the rate of change of slides is faster than flipping of pages when writing is done on it. Since a writing hand sequence corresponds to a slower

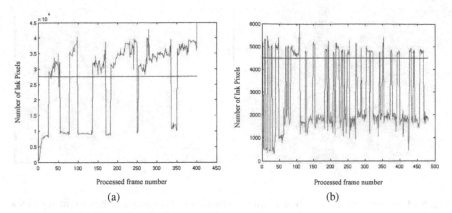

Fig. 5.9 Plot of the pedagogic content (number of ink pixels) of the processed frames for a lecture video containing (a) hand-written pages and (b) slide show (electronic slides). Note that the frame rate is reduced to 0.5 per second as explained in section 5.2.

activity, one may notice that the measure of content slowly increases with frame numbers and then reduces drastically when the page is replaced by a new blank page on which the instructor starts writing again. This is quite visible in Fig. 5.9(a).

The key-frame extraction algorithm, applied to a lecture video segment with handwritten slides yielded the key-frames as shown in Fig. 5.10. As it can be seen, there are 2 key-frames selected which represent around 10 minutes of the content segments in the video. As a comparative study, the same algorithm, without any skew correction method yielded duplicate key-frames some of which are shown in Fig. 5.11. The duplicate key-frames shown in Fig. 5.11 result from the translation or rotation (or both) of the page, when the instructor is writing and explaining the content. Note that both the frames in Fig. 5.11 have some positive skew angles with respect to the reference frames shown in Fig. 5.10, according to the sign convention given in Fig. 5.3. Now we analyze how the Radon transform eliminates these dupli-cate key-frames. The Radon transforms of the key-frames given in Fig. 5.10(a) and (b) in the reference direction are shown in Fig. 5.12(a) and (c), respectively. The 2-D Radon transform arrays computed for the skew frames shown in Fig. 5.11 (a) and (b) for the angular span as discussed in section 5.3 are shown in Fig. 5.12(b) and (d), respectively. The correlation of Fig. 5.12(a) with (b) and that of (c) with (d) are shown in Fig. 5.13(a) and (b), respectively. From Fig. 5.13, it can be noted that Fig. 5.11(a) has a skew of 1° and (b) has a skew of 5° with respect to their reference frames shown in Fig. 5.10. Hence these duplicate key-frames are discarded due to the high similarity of Radon transforms with the reference one, in a particular direc-tion. Hence the Radon transform based redundancy removal phase has effectively eliminated those duplicate key-frames which originate from the skew of the slide and thus yielded pedagogically unique or distinct key-frames as shown in Fig. 5.10.

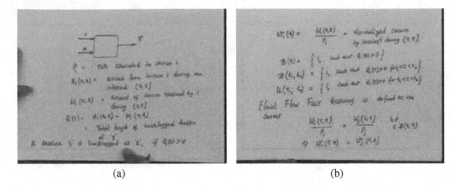

Fig. 5.10 Key-frames obtained from a lecture video segment consisting of hand-written slides as contents within a time window of 10 minutes. (a) and (b) are the two detected key-frames.

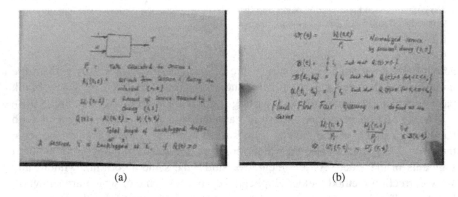

Fig. 5.11 Additional candidate key frames obtained from the same lecture video segment as in Fig. 5.10 with hand written slides as contents when no skew correction is employed. These duplications are removed and only key frames shown in Fig. 5.10 are retained.

Our algorithm, applied to another lecture video with slide shows (electronic slides) resulted in selecting 6 key-frames, as shown in Fig. 5.14, over a duration of 10 minutes. As the electronic slide presentation is faster than the writing activity, we can observe that the number of key-frames selected for the former will always be higher than that of the latter for a given time duration. These key-frames give the content description of the corresponding segments in a concise, but effective form. We observe that typically 2-8 key frames are selected over a 15 minute time window in the middle of the lecture hour to represent the pedagogic contents in the video. Interestingly, at the beginning of the lecture, this number is often less, signifying that the instructor spends some time in recapitulating contents of past lectures or expanding the motivation of a topic.

These selected low-resolution key frames are then processed in their temporal neighborhoods to yield super-resolved key-frames, a pair of which are given in Fig. 5.15. These are the corresponding super-resolved key frames of those low res-

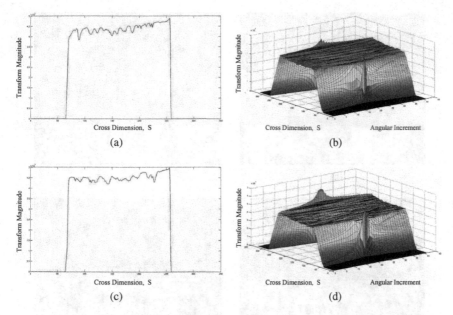

Fig. 5.12 Plots of Radon transform of content frames. (a) and (c) are the Radon transform of the content frames in Fig. 5.10(a) and (b), respectively in the reference direction and (b) and (d) are the 2-D Radon transform arrays in the angular neighborhood of the reference direction for the content frames in Fig. 5.11(a) and (b), respectively.

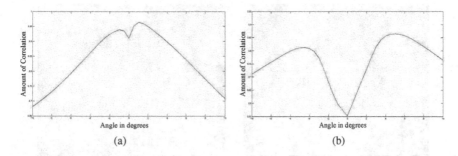

Fig. 5.13 Plot of the correlation of the Radon transform in the reference direction with those in angular span of ± 10 degrees. (a) gives the correlation between Fig. 5.12(a) and (b) and (b) gives that of Fig. 5.12(c) and (d).

olution ones shown in Fig. 5.10. The comparison of figures 5.10 and 5.15 demonstrates that the corresponding hand-written slides are much more legible when the super-resolution technique is used. In case of electronic slides, super-resolution is not needed if they are available in the original resolution. However if the electronic slide presentation is captured by the video camera and no source slides are available, super-resolution could be employed to enhance the resulting key-frames. A pair of such super-resolved key-frames are shown in Fig. 5.16 for slide show cate-

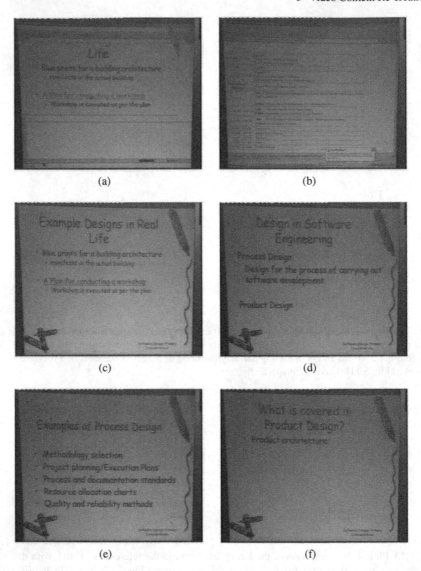

Fig. 5.14 All key-frames extracted from a lecture video consisting of slide shows within a time window of 10 minutes for a given lecture. Note that the number of resulting key frames are higher compared to Fig. 5.10, since electronic slides change at a faster rate, compared to hand-written slides.

gory. These are the corresponding super-resolved key frames of those low resolution ones shown in Fig. 5.14 (d) and (e).

In the case of non-content segments, i.e., talking head sequence, the key-frame extraction is done using the no-reference visual quality metric suggested in [154]. Fig 5.17 shows a sample result of detecting the best quality frame within the shot

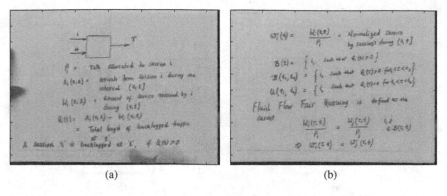

(a)   (b)

Fig. 5.15 Super-resolved key-frames of those hand-written slides shown in Fig. 5.10 for improved legibility.

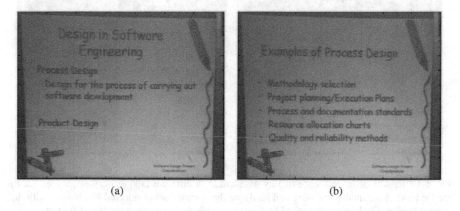

(a)   (b)

Fig. 5.16 Super-resolved key-frames of those electronic slides shown in Fig. 5.14 (d) and (e).

segment containing the talking head sequence. In Fig. 5.17(a) and (b) we plot the no-reference quality measures for all frames for two different instructors, and in Fig. 5.17(c) and (d) we display the frames corresponding to the highest quality measure. The parameters of the quality score evaluation, as given in equation (5.19), obtained with all test images are $a = 245.9, b = 261.9, c_1 = 0.0240, c_2 = 0.0160$ and $c_3 = 0.0064$ following the recommendation in [154]. The resulting best quality frames are used as key frames for the corresponding video segments.

This completes the preparation of the IMP which comprises of the super-resolved key-frames for content segments, key frames for talking head segments, the associated complete audio file and the meta-data file for the temporal information of key-frames. Hence the repackaging is effected by converting the highly redundant original video into a compact representation of its information content, namely IMP. The details of IMP are given in Appendix A. On implementation using Matlab 7.8, running on a machine with Intel Quad core processor and 4 GB of RAM, it took around 40 minutes to process a 1 hour video to generate its IMP. Though the HPP

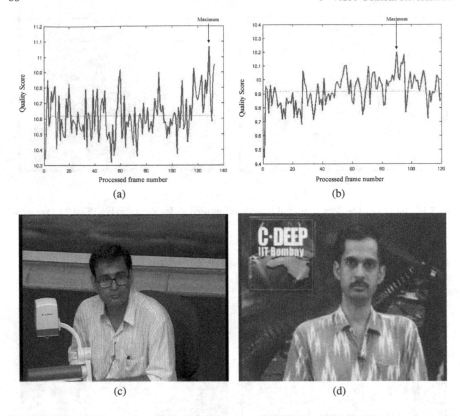

Fig. 5.17 Results of no-reference quality assessment in different talking head sequences. (a, b) are the plots of the quality measure and (c, d) are the corresponding selected key frames with the highest quality. The horizontal dashed lines in (a, b) indicate the average quality of frames.

based key-frame selection is fast, the increase in processing time is due to the in-built modules for skew correction and super-resolution for content frames.

The media is effectively re-created from the IMP, without any change in ped-agogic content, using FLASH. This 1 hour duration re-created *video* replaces the original data. The overall gain is that we were able to remove the redundant data in the content and non-content segments of the lecture video. As the repackaged media does not include any video segment, i.e., since the media is a synchronized play out of static key frames in relation to audio stream, the compression factor is very high. It could be as high as 50 for a slow paced instruction but we found that typically it lies in the range 10-25 over and above the MPEG coding scheme of the original video. The exact factor depends on the nature of the original instructional video in terms of how often the instructor is changing slides or pages of instruction. For ex-ample, in one set of results, we come up with a repackaged size of 52.4 MB for an original MPEG video of size 595 MB, thus yielding a compression factor of 11.35. For another MPEG video of size 1.3 GB, the repackaged media occupies 64.2 MB,

Table 5.1 Typical compression factors obtained through repackaging of different lecture videos

Video of 16 Min.	Original size	Size of IMP	Compression factor
Algorithms	586 MB	13.6 MB	43.08
Optimization	622 MB	13.9 MB	44.74
Random Processes	218 MB	13.8 MB	15.79
Image Processing	628 MB	13.9 MB	45.17

Table 5.2 Compression factors obtained for those lecture videos given in Table 5.1 using a specific audio codec, namely Samsung voice pen (SVR-S1540).

Video of 16 Min.	Original size	Size of IMP	Compression factor
Algorithms	586 MB	589.9 KB	993.39
Optimization	622 MB	730.8 KB	851.12
Random Processes	218 MB	631.9 KB	344.99
Image Processing	628 MB	712.6 KB	881.28

giving a compression factor of 20.23. These results are for one hour videos. Typical compression factors obtained for 16 minute videos are given in Table 5.1.

All the videos except *Random Processes* in the Table 5.1 are DVD quality ones. Note that the DVD quality videos yielded compression factors beyond 40. However, the non-DVD quality video was able to provide a compression factor of only about 15 since the eliminated redundant data (image frames) are less in size. This is because of the fact that it is the audio data which consumes a major portion of the IMP while it is the image frames which show such a spread in the case of a video. Hence we can note that for a given number of eliminated video frames, a larger frame size causes higher compression factors. Another fact that should be mentioned here is that the results given in the Table 5.1 are actually obtained by using the audio file in the MP3 format. One can note that even higher degree of compression could be achieved by employing a proper audio codec. As a study, we used Samsung voice pen (SVR-S1540) to record the instructional audio and found that the compression factor for the associated audio file format (SVD) is of the order of 1000. In such a scenario, the figure of merit of the suggested method of IMP creation is extremely high. The improved compression factors corresponding to those lecture videos given in Table 5.1 is given in Table 5.2 for comparison purposes. The effectiveness of our technique could be measured from the compression factors obtained. From the size of IMP given in Table 5.2 we observe that any instruction can be streamed to a client device at a rate lower than one kilo bytes per second. Hence streaming through a wireless medium could be achieved even at a remote place without any access to 3G technology. At the other end, the visual performance evaluation is highly subjective, and left as a future work.

In order to improve the situational awareness, one may prefer a virtual presence of the talking head (the part of video segment depicting the instructor teaching in the class room), instead of displaying the static frames of the talking head. We may insert a short video segment of the talking head at a regular interval in the repackaged media. The proposed system does allow the above insertion. However, the compression factor reduces to about 5-10.

5.8 Discussions

This chapter presented the development of a simple but accurate algorithm for a compression efficient and content preserved repackaging of lecture video sequences. In this, visual quality and content based key-frame extraction strategies are applied differently on separate classes of video segments and a pedagogic representation of these lecture video segments are done effectively. The no-reference visual quality metric by [154] was used for the key-frame selection of non-content segments like talking head while for content segments, an HPP based scheme is employed. The Radon transform based skew detection module avoids content key-frame duplication and yields distinct key-frames. The key-frames for different scenes of the lecture video provide a visual summary and an effective description of the content of the lecture. If the content key-frames suffer from poor resolution, they are super-resolved for improved legibility. On media synthesis, an estimate of the original lecture video sequence is re-created from these content key-frames, the talking head frame and the associated audio.

It can be seen that the reproduced media completely represents the original video in terms of all pedagogic values since it was delivered to the viewer in a semantically organized way, without altering the playback duration of the video. The multimedia summary produced by the content analysis which essentially contains a few video frames, a text file with the temporal information and the associated audio can be called an *instructional media package* (IMP).

The IMP is to be supplied to the users of distance education, so as to effect the content dissemination of lecture video. The users, normally students, view the lecture *video* by a proper content re-creation mechanism. The summarization and the on-demand content re-creation of lecture videos lead to considerable savings in memory when an enormous amount of video data are to be stored. This, in turn, accounts for a fast browsing environment, on a huge volume of lecture video data. Another achievement that can be noted is, the reduction in the transmission bandwidth. The next chapter brings about an extension of this method of video content re-creation which is meant for the display of IMP on miniature devices like mobile phones and PDAs.

Chapter 6
Display of Media on Miniature Devices

6.1 Introduction

The increased use of hand held devices and the significant development of mobile multimedia technologies may soon usher in a paradigm shift in learning means from desktop to outdoor which inevitably requires viewing video on miniature displays. Being intended for display on larger screens like television and computer monitors, the original videos when displayed simply by resizing it to fit to the smaller screen, some crucial parts of the visual content become too small to be legible or even lost. In this scenario, we come up with a novel method of retargeting lecture video on mobile devices by which the original lecture video is adapted in a highly reduced form but with better means to suit the target display and storage with the minimal information loss.

Small displays can deliver less content to the viewer and hence our focus is on an efficient visual delivery by which only a selected part of the entire video frame that is relevant at the instant of video playback is displayed at its fullest resolution. That is, the degradation in the legibility of the displayed frame due to the limitations on screen size can be effectively overcome by delivering a smaller region in the video frame that is equal to the display resolution of the mobile screen. Then a student would be really benefited by the fact that he/she can watch the content of the video any time with the maximum legibility on the miniature screen. This is the motivation behind this work. The key idea is that a virtual panning of the original key-frame with a window size equal to the size of the miniature screen in accordance with a meta-data created from the writing sequence in the original video would suffice to render the maximum legibility to the viewer. A related work was done in [79] through which the authors propose a retargeting method for conventional video with automatic pan using motion information from the original video. The algorithm crops all the frames according to the local importance of information so as to fit it on the target display. Hence a spatially cropped video is transmitted instead of just the key-frame used in this study. In another work [158] the retargeting is effected by the shrinkage of less important regions in a commercial video which are identified by local saliency. In [128] the authors adopt a domain specific approach to exploit the attributes of video for an intelligent display of soccer video on mobile devices. We can find that most of the retargeting methods found in literature for conventional videos [62, 73, 130] use cropping, scaling, seam-carving, and warping to effect the desired adaptation on miniature devices. Since the object motion is less in instructional scenes, these methods are not suitable for retargeting of lecture videos on miniature mobile devices. Video transcoding is another technique which is related to this work. Since transcoding methods [1, 161, 166] tend to decrease the display resolution when the media is transmitted or streamed to a miniature device, they also do not hold good for the adaptation of lecture video on miniature devices. Hence we adopt a totally different retargeting approach in which the contents of the *key-frames* extracted from video are delivered to the viewer by using a *moving key-hole image* (KHI) through animation in the mobile device itself. We also offer a multi resolution intelligent visual delivery based on the text tracking meta-data obtained from the original video.

For the extraction of key-frames from the video, we have to first segment and then detect the activities in it. Then the key-frames in content and non-content segments of the video are identified by using appropriate techniques. For all the procedures like temporal segmentation, shot recognition and key frame selection we adopt the techniques explained in previous chapters. The instructional media package (IMP) derived from an original video is to be finally retargeted on miniature devices. Thus this work of the legibility retentive display scheme of IMP on miniature devices is an extension of the work propounded in Chapter 5 of this book.

On media re-creation, we have to perform animation of these key frames for the corresponding time span in the original video along with the audio stream in complete synchronization. When the content key-frames are displayed on a smaller display like a mobile phone screen, some information might be inherently lost either by spatial sub-sampling or by marginal cropping in an attempt to fit it on the screen. Hence we propose a tracking meta-data based *virtual panning* of the content key-frames with a window size equal to the size of the mobile screen by which only the *region of interest* (ROI) in the content key-frame is delivered to the viewer with its maximum resolution. This meta-data is to be created at the server side by analyzing the writing succession in the content key-frames. Tracking the pen in the frame sequence could be one of the methods of creating this meta-data. For example in [91] the authors use a combination of a correlation detector, Canny edge detector and Kalman filter for tracking the trajectory of the tip of the pen. Since their objective is an accurate character recognition, these sophisticated methods are essential. The same method with minor modifications is used for video based pen tracking in [39] through which also the authors aim at handwriting recognition. Since our goal is to just select a region of interest in which the pen currently strikes, these complicated methods are not at all necessary. Further, during the discussion in the class room the instructor often moves his/her pen away from the paper for a prolonged duration and then starts writing again. The above method fails to track the pen when these digressions take place during the lecture. We develop a simpler and fast method to detect the region being scribbled and its center point is taken as the x-y co-ordinates

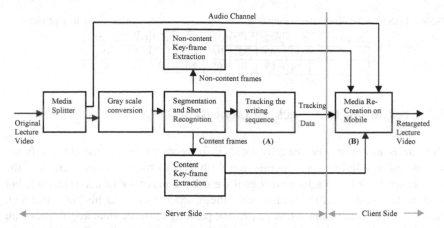

Fig. 6.1 A block schematic for the suggested method of legibility retentive display of instructional media on portable devices.

of the tracked text. Since the projection profile is well known for its ability to resolve document content in x and y directions, we propose to use the HPP and vertical projection profile (VPP) of the ink pixels in the content frames for the tracking of written text.

The talking head key-frame, the content key-frames together with the tracking meta-data and the associated audio form an effective repackaged media on its deployment at portable devices. Hence our algorithm aims at the summarization of lecture video in terms of key-frames, the related text tracking meta-data and audio at the server side and at the client side, the re-creation of the video using a programmed playback of these data so as to fit the miniature display in a better way, yet preserving its pedagogic values.

Our objective here is the preservation of legibility of the written text during the retargeting of instructional media on mobile devices. For this, we come up with a complete system, the block schematic of which is given in Fig. 6.1. It works with the following distinct steps for the repackaging of the lecture video:

(a) Shot detection and recognition
(b) Key-frame extraction for content frames
(c) Key-frame extraction for non-content frames
(d) Creation of text tracking meta-data, and
(e) Legibility retentive media re-creation on mobile device.

Since steps (a) - (c) have already been discussed in previous chapters of this book, they are not explained here and only the last two steps (shown as blocks A and B in Fig. 6.1) given above are discussed. These include a text tracker that creates the meta-data required for panning of the content key-frame during play back and a media re-creator on mobile device, that displays the media by an intelligent visual delivery without any loss in pedagogic content or effectiveness of the instructional video. These two steps are dealt with in detail in the following sections.

Table 6.1 Detected co-ordinates during text tracking in the content segment of a lecture video.

frame no.	...	6375	6425	6475	6525	6575	6625	6675	6725	...
x	...	34	63	82	109	138	187	208	246	...
y	...	84	89	70	68	85	88	82	90	...

6.2 Creation of text tracking meta-data

Our aim is to deliver the spatially local, visual content in the content key-frame with its full resolution to the mobile user without affecting the effectiveness of the instruction. In order to achieve this goal we use the playback of audio as such, but focus at that portion of the frame where the instructor explains his/her scribbling. Hence we approach this problem as a virtual panning of the content key-frames with a selection window whose size is equal to that of the resolution of mobile screen and whose movement is automated through the text tracking data obtained from the original video. It requires meta-data creation on the server side by which the co-ordinates required for positioning the selection window on client side are generated from those of the writing sequence on content frames. This meta-data essentially contains the x-y co-ordinates of the currently written point on the writing pad and on the temporal basis, it is an array of size $2 \times L$ where L is the total number of frames in the content segment of video data. Note that L is typically much higher than the count of frames in non-content segments since writing operation is a much slower activity compared to the standard video frame rate, say, 25 fps. Table 6.1 represents a typical co-ordinate array for tracking a handwritten slide sequence. The x-y co-ordinate in the table represents the center point of the selection window (ROI). Here the selected sequence contains 350 frames and the co-ordinate array representing the tracking of text is computed in the interval of 50 frames. That is, since the writing activity is slow, we track the written text in the interval of 2 seconds while creating the meta-data. The corresponding task that should be performed at the client side will be explained later in this section.

As a general practice, instructors start writing from the top of a fresh page and end up at the bottom. Hence the textual content in the frame is produced as a result of writing succession by the instructor from the top to bottom of the slide. Occasionally he/she can scribble in between also. It takes thousands of frames for a slide to get completed and the lines of text (in turn, the text in a line also) are assumed to be produced in a sequential manner. Hence one may adopt the following strategy to detect the position of writing at a given instant of time - (i) first detect the line (y-position) along which the instructor is writing and (ii) then detect the x-position of writing in that line. In this study we explain the use of projection profiles of the binarized content frames for the text tracking. We first detect the y-position of the currently written text using a comparison of the ink pixels in the HPPs of the consecutive content frames and then determine the x-position from the VPPs of the ink pixels contained in the cropped regions around the y-position of text in these frames. Remembering that HPP $F(m)$ of an ink pixel detected frame, defined by equation (5.7) in Section 5.2, is an array of M elements, we employ it to compute the vertical position of the text, that is being scribbled at a given time instant. To detect the y-position of the currently written text, we locally count the amount of differential ink pixels in the HPPs of two consecutive frames and locate the point where it grows to a considerable value. For this we subtract the HPP of the previous frame from that of the current one to get a difference HPP array

$$D_H(m) = F_t(m) - F_{t-1}(m).$$ (6.1)

Now local summation of the absolute difference HPP array $D_H(m)$ is performed to detect any considerable variation in ink pixels. Hence we calculate

$$SD_H(j) = \sum_{m=j-w_d}^{j+w_d} |D_H(m)|; \, w_d < j \leq M - w_d$$ (6.2)

and if $SD_H(j) > T_D$ for $j = l$, we assume that there is a scribbling activity in the lth line since the local sum of the absolute HPP difference yields a high value which is a result of the introduction of extra ink pixels in that line. Here w_d is the localization parameter for summation and T_D is the threshold for differential ink detection. It could be noted that with $w_d = 0$, equation (6.2) yields just the absolute difference of the HPPs for a particular line $j = m$. This is found to be insufficient for an accurate detection of the increase in ink pixels and would result in false alarms. Hence in order to increase the information content, we integrate the information $|D_H(m)|$ locally around the line j before detection. Considering the average size of the fonts in handwritten texts, we choose $w_d = 10$. Hence the local summation is for ± 10 lines around a line $j = m$. This is checked against the given threshold to detect writing activity in that line. Hence by using HPP we get the y-position of the currently written text which is one component of the tracking meta-data. The next task is to locate the x-position of the currently written text. This can be obtained by creating the vertical projection profile (VPP) of the ink pixels around the readily available y-position. So we crop horizontal strip images around the y-position for the current and previous frames and take the difference VPP of these cropped images

$$D_V(n) = V_t(n) - V_{t-1}(n)$$ (6.3)

where $V(n)$ is the VPP computed for the strip image from the binarized content frame $B(m,n)$ by a similar procedure as given in Section5.2. $2W$ is the width of the strip image such that $l - W < m \leq l + W$. We choose $W = 10$ in this study. Now local summation of the absolute difference VPP is performed to detect the x-position of the written text. Hence we compute

$$SD_V(k) = \sum_{n=k-w_d}^{k+w_d} |D_V(n)|; \, w_d < k \leq N - w_d.$$ (6.4)

The point $(k = i)$ at which this local sum yields the maximum value indicates the introduction of new ink pixels along that specific horizontal line, l. This point is

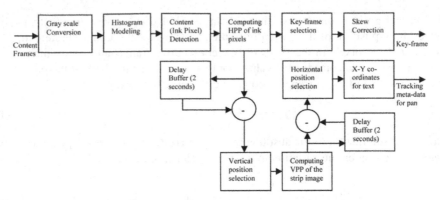

Fig. 6.2 Block diagram illustrating the extraction of tracking meta-data and key-frames from content frames.

taken as the x-position of the written text. Now the tracked point is (l,i) which denotes the center co-ordinates of the ROI. Hence we use the projection profile of ink pixels in two perpendicular directions to derive the tracking meta-data for the newly introduced text.

Since the writing pace of the instructor is very slow compared to the video frame rate, the increment in content (extra ink pixels being introduced) from frame to frame is too less to be of noticeable difference in HPPs. As a matter of fact, we look for the increment in content during the interval of 2 seconds for further processing of the information about written text. This sub-sampling of frames by a factor of 50 not only increases the efficiency of our algorithm, but also reduces unnecessary computations. Such a scheme has already been used in Section 5.2. Our assumption is justified by the fact that there is not much variation of textual content within a 2 second duration in a lecture video. The block schematic showing the extraction of tracking meta-data along with content key-frames is shown in Fig. 6.2. The content frames are first converted to gray-level images and from the intensity histogram, an optimum threshold for ink pixel detection is found out as explained in Section 5.2.1. Then HPPs of the ink pixel detected frames are constructed and the difference HPP in an offset of 50 frames (2 seconds delay) yields the y-position of the written text as per equation (6.2). From the VPP of the horizontal strip images around this y-position, we compute the x-position of the written text according to equation (6.4). The resulting x-y co-ordinates are stored in a meta-data file so as to be deployed on mobile devices along with the key-frames and the audio for content re-creation.

The tracking data derived in regular intervals of 50 frames cannot be employed as such to drive the panning window at the media re-creator since it causes a jittered panning. Hence to make it smooth, we linearly interpolate the data points in between t^{th} and $t + 50^{th}$ frame positions as shown in Fig. 6.3. The interpolation can be done directly by the client at the time of playback, saving in meta-data size. Alternately, it can be generated at the server side. In between two data points $(x(t_1),y(t_1))$ and $(x(t_2),y(t_2))$, the location of the new window at time t is defined as

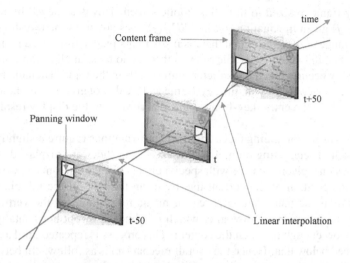

Fig. 6.3 Creation of tracking meta-data in content frames. Note that the meta-data is obtained in the intervals of 50 frames and, in between, they are interpolated.

$$x(t) = \frac{x(t_2) - x(t_1)}{t_2 - t_1}(t - t_1) + x(t_1); \ for \ t_1 < t < t_2. \tag{6.5}$$

Similarly we interpolate for $y(t)$ also. Note that we use $t_2 - t_1 = 50$ in this study. This interpolation of tracking data between the selected frame positions results in a smooth movement of the panning window across the content key-frame. This video meta-data is provided along with the key-frames and the audio when it is deployed on the mobile display unit for *virtual* playback. All the information like name and type of frames, their temporal validity, text tracking data, name of the audio file, etc., required for the playback of the media on mobile devices is stored in an **XML** file and is kept in the same folder as the image and the audio files. This folder containing the *instructional media package* (IMP) will serve as a source for content re-creation at the client device. The details of IMP are included in Appendix A.

We observe that the instructor, at times, moves his/her pen position up and down at a fast pace to relate to various parts of the instruction delivered till that point of time. It is not advisable to pan the image at a fast rate when the viewer is not able to follow the instruction. We incorporate a provision for varying the size of the selection window to increase the region of interest under extreme movement of the writing hand in the content frame. This is done automatically using the variation of the tracking data from frame to frame as explained in the next subsection. Another mechanism for an optional manual control for the movement of the selection window on the content key-frame is also included during the implementation of our algorithm. This is to enable the viewer to manually move the region of interest as he/she wishes since the tracking data may fail occasionally. Moreover, if the content frame has a prolonged *idle hand* or it is a *clean frame* for sometime, we display the whole key-frame, resized to fit to the mobile screen. This is achieved by using the term $SD_H(j)$ given in equation (6.2). If $SD_H(j)$ does not cross the threshold T_D for the entire range of j, we infer that there is no extra ink pixels introduced in the frame due to the fact that the hand is idle or even there is no hand at all in the frame. This situation may occur when the instructor digresses from the topic. Then the region of interest does not conform with the key-frame and we do not go for a key-hole image, instead we show the entire key-frame resized according to the display resolution of the mobile deice.

Lecture videos containing electronic slide presentations require a slightly different approach of retargeting on mobile devices. If the slides are displayed on incremental lines on a given page, or with special effects like text streaming, we are able to detect the position of the current line by using the HPP of the ink pixels. This would serve as the meta-data required on media re-creation for the vertical positioning of the selection window after which it is linearly swept horizontally with a time-line slow enough to watch the content. This process is repeated until a new line is introduced below that. Hence the overall mechanism is as follows. In between the occasional y-increments, the selection window will be swept a few times along the x-direction. If the slide show is performed with the cursor to explain its content, its tracking will provide the required meta-data for positioning the selection window. But a mere vocally referred slide or a laser-pointed slide presentation fails to produce any meta-data at all from the video frames. Here the only thing that can be done is to slowly pan the selection window a few times horizontally and then step vertically to repeat the same through another line. The overall process is repeated several times to scan the entire key-frame. The vertical step size can be calculated from the ratio of the vertical dimensions of the display of the server and the client device. Then the time interval for the sweep along a particular line can be calculated from the total time duration required for displaying that key-frame and the vertical step size. However, we find that a manual panning of the display works better for such kind of slide presentation.

6.3 Legibility retentive media re-creation

Finally all the key-frames, together with the tracking meta-data and the associated audio, should provide a legibility retentive visual delivery on the mobile device. The deployed data in the mobile device consist of the following : (a) few key-frames in any image file format, (b) an audio file in a suitable file format and (c) an XML file containing the temporal marking for the placement of key-frames and the meta-data for the automatic panning of the content key-frame. We note that a key-frame which is the maximally filled content page of the writing hand or slide show sequence was detected after having completed writing the page by the instructor, while the discussion about the content of the page had started just after completing the last working page. Hence during media re-creation, this detected key frame is placed at the beginning of the corresponding time slot since the selection window is programmed

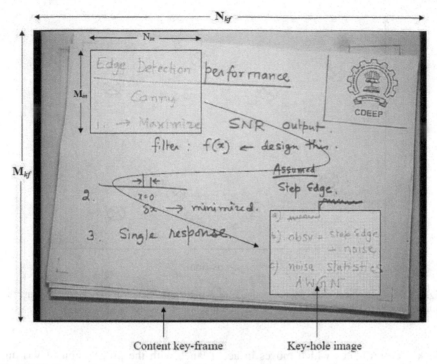

Fig. 6.4 Delivering the visual contents of a key-frame through a moving key-hole image in accordance with the tracking data. $M_{kf} \times N_{kf}$ is the size of the key frame and $M_m \times N_m$ is the size of the mobile screen. The curve shows the progression of the key-hole.

to scan the entire content of the frame which should be fully available at the time of audio playback.

Since media re-creation is done on a mobile device, the choice of the device and the corresponding platform offers an implementational diversity. In this work we give an implementational example with windows mobile platform [156], while an online demonstration is provided on an android platform. We design a mobile multimedia player called *Lec-to-Mobile*[1] (Lecture to Mobile) for porting instructional media on mobile devices. Its design aspects are given in Appendix B. The representative key-frame of the talking head sequence is just scaled down and displayed on the mobile screen as they account only for some sort of situational awareness to the viewer. For content segments like writing hand, if the key-frames are displayed using downsampling, the viewer will not be able to read the text on the content frame. Here the written content is to be delivered to the viewer with the maximal legibility. This is achieved by displaying only a selected region on the key-frame in which the selection window size is equal to the resolution of mobile screen and by controlling its movement over the original key-frames using the text tracking data obtained from the original video as shown in Fig. 6.4. This tracking data denotes

---

[1] Application for trademark has been filed by Indian Institute of Technology Bombay.

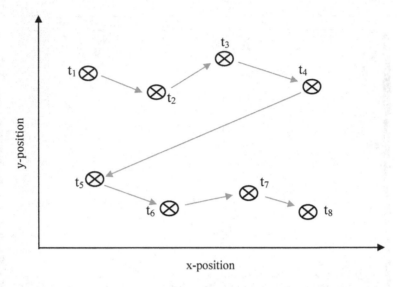

Fig. 6.5 Plot of typical track points obtained for a particular content frame. Here t_k means the instant when $(x(t_k), y(t_k))$ represent the corresponding window center.

the x-y co-ordinates which moves in accordance with the progression of writing from frame to frame and within the key-frame. Hence the written content is delivered in its maximum possible resolution to the viewer by considering the contextual relevance of the text regions in the content key-frame at various instants of instructional activity. Hence for saving memory we exploit the temporal redundancies in the original video and for an efficient visual content delivery of the key frame, we employ a contextual relevance based approach. Note that the relevant textual region in the key-frame at any instant of time is also being referred in the audio track of the instructor. In other words, our algorithm produces a sequence of a few key-frames from the original video and then displays them as a moving key-hole image, as directed by the tracking meta-data. On playback at the client site, this moving key-hole image along with the audio form a fully re-created media with improved visibility on the miniature screen. The movement of the key-hole image over one particular key-frame is for the time duration for which that key-frame is a representative one for that entire pedagogic video paragraph.

If there is a considerable deviation in the tracking co-ordinates from instant to instant, we infer that the instructor is toggling from one point in the paper to another which is beyond both the spatial span and movement of the selection window. An example set of track points is shown in Fig. 6.5 for illustration. Here the panning of the selection window is slow from t_1 to t_2, t_3 and t_4 but fast from t_4 to t_5 after which again it is slow through t_6, t_7 and t_8 . During the interval of the fast movement, we purposefully increase the size of the selection window (by downscaling the picture appropriately) for wider visibility such that the viewer does not miss any region of interest and to provide a sense of spatial context. Hence between t_4 and t_5 we adopt

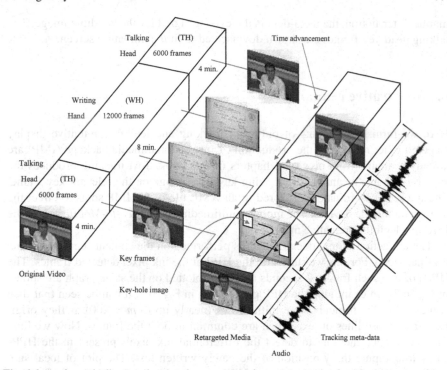

Fig. 6.6 A schematic diagram showing the sequence of an original video for 16 minutes, resulting key frames and the re-created media, along with the audio and the meta-data for displaying the content key-frames as a moving key-hole image. Note that no tracking meta-data is required for the non-content frames.

a selection window of higher size large enough to include position $(x(t_4), y(t_4))$ and $(x(t_5), y(t_5))$ totally inside the display. Under these circumstances the text does not appear in its full resolution to the viewer since scaling is required to fit it on the screen, and the text may not be properly legible during this time interval. As explained earlier, a provision for displaying the entire key-frame for the time duration for which the instructor digresses from the topic written on the slide is also incorporated when the entire media is displayed on the mobile device by appropriately downsampling it to fit the display resolution. Another important feature of media playback is a manual over-ride control. By this, a viewer can choose his/her own region of interest over the displayed key-frame. This mechanism may be useful on playback since the tracking meta-data may occasionally fail to display the vocally referred region on the key-frame. This manual over-ride mechanism provides an interactive control, in addition to the normal automatic playback of the IMP. A diagram illustrating the overall methodology for a 16 minute video is given in Fig. 6.6. Here there are four extracted key-frames of which two are content ones and the other two are not. The tracking meta-data is available only during the content key-frame display by which the key-hole image of size equal to that of the resolution of the miniature screen moves across the frame as shown in the figure. Unless there is a manual interruption, the meta-data is the driving signal for the key-hole image. The talking head key-frames are shown downscaled to fit the miniature screen.

6.4 Illustrative examples

Here we furnish the results of only text tracking and legibility retentive display since all other results for the production of instructional media package (IMP) are already available from previous chapters of this book. Text tracking produces the x-y co-ordinates meant for the moving key-hole image on the relevant key-frame. The tracking co-ordinates are stored in an XML file of the IMP so as to effect the display on mobile devices. The mobile multimedia player, *Lec-to-Mobile*[2] plays the IMP in a legibility retentive manner.

The extraction of text tracking data is performed on the content segments of the original video. For this we start with the HPP of the ink pixel detected frames. The HPPs of two such frames 2 seconds apart and plotted on the same graph are shown in Fig. 6.7. On examining the HPP plots given in Fig. 6.7, it can be seen that they coincide locally, through some initial rows (nearly up to $m = 200$) as they originate from those lines of text which are common in both the frames. Now we look for a difference measure to detect the differential ink pixels present in the HPPs so as to compute the y-position of the newly written text. The plot of local sum of the absolute difference of the two HPPs given in Fig. 6.7 is shown in Fig. 6.8. The point where a considerable deviation starts is noted as the y-position for tracking. From the plot we note that this is for $m = 222$ which is actually the tracked y-position of the written text. The VPP based x-position extraction method is illustrated in Fig. 6.9. The cropped horizontal strip region around the y-position for the two frames are shown highlighted in Fig. 6.9(a) and (c), and the corresponding VPPs are shown in Fig. 6.9(b) and (d), respectively. The plot of local sum of the absolute difference of these VPPs is shown in Fig. 6.9(f). The region where a high overshoot occurs corresponds to the highlighted rectangular area in Fig. 6.9(e) which essentially contains the newly written text. Hence the point corresponding to the maximum of the curve in Fig. 6.9(f) is selected as the x-position to track the text. From the plot, it is seen to be $n = 400$ and hence at this given instant the tracked point is $(222, 400)$ in the key-frame. This procedure is repeated for all content video segments in the intervals of 50 frames. The corresponding video meta-data along with key-frames and audio are fed to the mobile device for media re-creation.

On media re-creation the representative key-frame of the talking head sequence is scaled to fit the resolution of the mobile screen and displayed for the entire time duration for which the corresponding segment appears in the original video. The same method as applied to content key-frames will result in poor legibility as shown

---

[2] An android based mobile multimedia player called Lec-to-Mobile is freely downloadable from the site https://market.android.com/details?id=com.lectomobile using the mobile phone. Alternately, it may be downloaded from http://www.ee.iitb.ac.in/~lec-to-mobile. Some sample media which can be played by Lec-to-Mobile are also available at the IITB site.

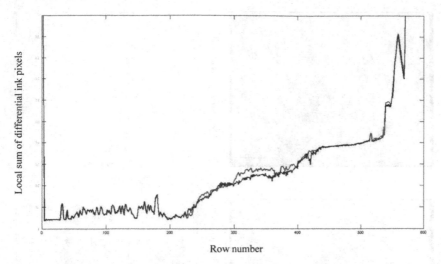

Fig. 6.7 Plot of horizontal projection profiles (HPP) of two temporally adjacent frames for a lecture video containing hand-written pages. Note that the plots almost coincide for the rows nearly up to 200.

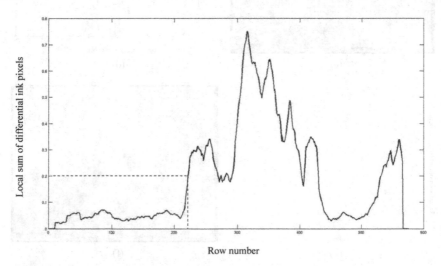

Fig. 6.8 Plot of the local sum of the absolute difference of the HPPs shown in Fig. 6.7. Note that for $m = 222$, the difference exceeds a given threshold.

in Fig. 6.10(a). We solve this problem by using the suggested automated key-hole imaging method in which the written content is delivered to the viewer with the maximal legibility but locally delimited as shown in Fig. 6.10(b). This cropped image is actually a child frame obtained from the parent key-frame at the instant of playback. As explained the text tracking meta-data drives the cropping window of

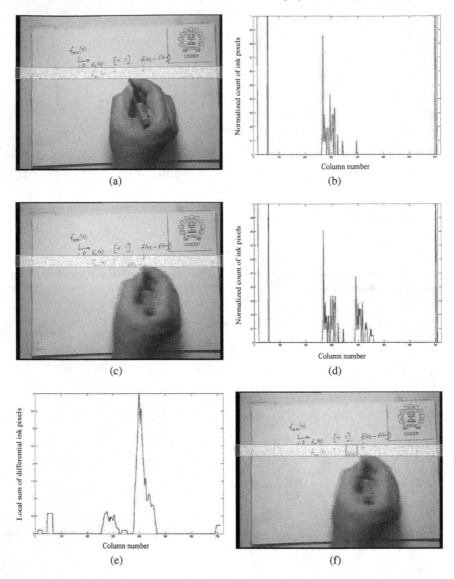

Fig. 6.9 Illustration of VPP based x-position tracking. (a) and (c) are a pair of processed frames in which the highlighted horizontal strips originate from the already tracked y-position of the text, (b) and (d) are the corresponding VPPs of the strips in (a) and (c), respectively, (e) shows the local sum of the absolute difference of the VPPs shown in (b) and (d), and (f) shows the (detected rectangular) region of the newly written text. Note that the difference peaks at $n = 400$.

size equal to that of mobile screen to scan the entire content in the key-frame for the entire duration for which that particular key-frame is intended for display.

(a)   (b)

Fig. 6.10 Comparison of performances of the suggested method and the existing technology. (a) Conventional display of video frame on a mobile screen in which scaling degrades the resolution and (b) the legibility retentive local, visual content delivery on the mobile screen using a virtual pan.

The media re-creator occasionally increases (say, doubles) the region of interest based on the amount of variation found in the tracking meta data. If the tracking meta data is compactly bounded, the size of the region of interest is constant throughout the play back. Otherwise we utilize a multilevel resolution based visual delivery which generates the feel of automatic *zoom in/out*. For this we calculate the distance, $d(t_1, t_2) = \sqrt{(x(t_2) - x(t_1))^2 + (y(t_2) - y(t_1))^2}$ where $(x(t_1), y(t_1))$ and $(x(t_2), y(t_2))$ are consecutive data points (refer to Fig. 6.5). If $d(t_1, t_2) > 0.5R$, we go for a higher size of the selection window which inherently results in a reduced resolution on video playback, as the mobile screen displays it appropriately resized. Here R stands for the display resolution of the miniature device. We select the smaller of the two dimensions (horizontal or vertical) to denote R. A similar situation would occur when we show the entire key-frame as the instructor deals with some topic other than what was written on the slide. The manual control over the region of interest is deactivated during the display of the entire content key-frame as no panning is possible.

Hence we see that the cropped display of the content key-frame maximizes the efficiency of visual content delivery to the mobile user. During an instructional shot, this cropping window moves over the key-frame to introduce *virtual* pans and zoom in/out, ensuring a kind of dynamism to the viewer. We demonstrate results of retargeting a variety of lecture videos to smaller screens of hand held devices through the online interface at android market.

6.5 Discussions

This chapter presented a novel method for legibility retentive display of instructional media on mobile devices that preserve the pedagogic content in the original instructional video but reduces storage requirements drastically. Thousands of video frames were represented by only a few key-frames which were displayed on the mobile screen as moving key-hole images. The size of the key-hole image was comparable with that of the mobile screen and so only the *region of interest* (ROI) in the key-frame with respect to the audio was displayed with the fullest possible resolution. Quite naturally, it is assumed that the audio is in synchronization with the pace of scribbling on the writing pad by the instructor, which is most often correct. The movement of the key-hole image (KHI) was controlled by a meta-data derived from the tracking of the written text in the content segments of the video. This meta-data was derived by using both the HPP and VPP of the ink pixels in the content frames. Originally these meta-data were computed at 50 frame intervals and during the display on mobile devices, these are interpolated to yield smooth panning at the 25 fps rate. We also make provision for animating the pan at a lower fps in the designed and freely available Lec-to-Mobile player in case the miniature device has less computing power.

Even though this method eliminates the temporally redundant frames in the content video segments, it does not crop the resident key-frames in the memory. Instead, it selects a rectangular region of interest in accordance with the tracking meta-data and delivers this region on the mobile screen on play back. In the default case, this panning window moves automatically with the tracking meta-data. Occasionally one can have a manual control through touch screen or the selection key on the mobile keypad. Thus the movement of the window can be additionally controlled manually as per wish of the viewer if for some reason the hand tracking meta-data performs poorly. Moreover, the local delivery of the visual content is not always performed with one single (fullest) resolution. If variation in the text tracking meta-data is much more than expected, the algorithm expands the region of interest when the visual delivery will be in a reduced resolution. Hence by using this technique, an intelligent visual content delivery can be achieved on miniature devices having a small display unit.

This legibility preserved retargeting of instructional video on miniature devices has a great potential of finding applications in distance education systems as now-a-days the end-users are also mobile. This adds to the goals of distance education for providing the learning means *anywhere, anytime*. Hence we see that the repackaged lecture video called instructional media package (IMP) may serve as a powerful learning material in distance education. At the same time, the providers of distance education, who are actually the owners (or license holders) of the original video should be able to assert their copyrights through proper information security systems. The next chapter discusses about the copyright protection of IMP through image and audio watermarking schemes. The watermarking of IMP is assumed to be done before their deployment on miniature devices.

Chapter 7
Copyright Protection

7.1 Introduction

The impact of digital consumer devices and the Internet on our day-to-day life has brought about a drastic progress in capturing, transmitting, storing and remixing of digital data. This, in turn, gives rise to challenging issues in data security to prevent its illegal use. In the area of distance education, the service providers should be able to protect the production of educational videos and the related materials from piracy. Hence the ownership rights should be preserved before these videos are deployed among students. As a mean to provide copyright protection of these data, the technique of digital watermarking may be employed. Watermarking is the process by which a digital code, i.e., the watermark[1] is imperceptibly embedded in the host media[2] such that the identity of the data owner (either in the form of a specified pseudo-random sequence or a given seal of the content owner) is hidden in it. The watermark characterizes a form of owner certificate applied on the host media and thus it marks the host media as being an intellectual property. The pre-registered watermark can be extracted by the creators to prove their ownership on the host media, whenever a question about the ownership arises. The host media may refer to images, audio or video. Since our objective here is the protection of the instructional media package (IMP), we aim at the digital watermarking of key-frames and audio separately.

A general schematic diagram showing a watermark life-cycle is shown in Fig. 7.1. The watermark (W) is inserted in the host media (M) using a key (K) which is known only to the owner. The watermarked media (M') which is distributed to the world may occasionally be subjected to *attacks*. An attack generally refers to the processing of the media (M') such that the watermark (W) becomes irretrievable. In the watermark extraction strategy, the watermark is retrieved from the given data using the abovesaid key (K). A similarity measure (R) between the extracted watermark

---

[1] Watermark is also called the message.
[2] Host media is also called the cover.

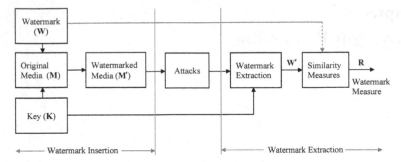

Fig. 7.1 A block schematic diagram of a generic watermark insertion and extraction system.

(W′) and the embedded watermark (W) is computed to prove the copyright. In order to be effective, a watermark should possess the following characteristics [29].

Unobtrusiveness: A watermark should be well hidden such that it may not interfere with the perception of the desired media. There is always a trade off between the extractability and imperceptibility of a watermark. Hence the watermark should be inserted by making a compromise between the two.

Robustness: Once a watermark is properly inserted, it should be immune to any attempt for its removal. That is, any effort to destroy the watermark so as to remove the seal of ownership protection should result in a severe degradation of the media. Specifically, the watermark should be robust against filtering, compression, geometric distortions (rotation, translation, cropping and scaling in the case of image and video), and subterfuge attacks like collusion and forgery.

Universality: One watermarking algorithm should hold valid for all the three media (image, audio and video) of interest. This property contributes to the potential application of joint watermarking of multimedia documents.

Unambiguousness: The owner should be unambiguously identified on the retrieval of the watermark. The accuracy of ownership identification depends on the depth of watermarking, which is directly opposed by the first characteristic, namely unobtrusiveness.

Watermark embedding can be performed in temporal (audio), spatial (image), spatio-temporal (video) or transform domain (audio, image or video). A multimedia document generally consists of a collection of images, audio files, video and optional text files. The IMP produced by the method explained in chapter 5 is such a document which is mainly comprised of the key-frames and audio, extracted from the original video. Hence watermarking schemes are to be applied in a suitable domain, to the constituent media of IMP. This essentially includes image watermarking and audio watermarking, which are explained in the following sections. Before entering into these topics, it is better to have a comparison of the two and also to introduce the idea of spread spectrum modulation (SSM) which is commonly employed for secure watermarking.

7.2 Watermarking methods for image and audio

The characteristic features that a watermark should possess, as given before, are common to both image and audio watermarking schemes. In [148] and [149] authors present a generalized watermarking framework that applies to both types of media. But there are still significant differences between these media watermarking schemes [10]. A watermarked image, being two dimensional, is susceptible to a vast variety of attacks like rotation, translation and scaling while a watermarked audio is devoid of such attacks. While audio watermarking methods usually work on either time or Fourier domain, image watermarking methods often use DCT or wavelet domain which is closely related to the JPEG compression standard. Also due to the difference in characteristics of human visual and auditory systems, different masking principles should be taken into account in each case. In general, the human auditory system (HAS) is more sensitive to disturbances than the visual system, which makes audio watermarking a very challenging task.

Digital image watermarking schemes can be broadly divided into two categories - spatial domain and transform domain techniques. The spatial domain methods include least-significant bit (LSB) [142], patchwork [164] and spread spectrum modulation (SSM) [143] based techniques. An advantage of the spatial domain techniques is that they can be easily applied to any image, regardless of subsequent processing (whether they survive this processing, however, is an entirely different matter). A possible disadvantage of spatial domain techniques is that they do not allow for the exploitation of the subsequent processing in order to increase the robustness of the watermark. Also, adaptive watermarking is a bit more difficult in the spatial domain. Both the robustness and quality of the watermark could be improved if the properties of the cover image could similarly be exploited. For instance, it is generally preferable to hide watermarking information in noisy regions and edges of images, rather than in smoother regions. This makes transform domain methods very attractive and hence they tend to be more popular compared to spatial domain methods.

In transform domain techniques, one aims at embedding the watermark in the spectral coefficients of the image. The most commonly used transforms are the discrete cosine transform (DCT), discrete Fourier transform (DFT), discrete wavelet transform (DWT), and the discrete Hadamard transform (DHT). The reasons for watermarking in the frequency domain is that the characteristics of the human visual system (HVS) are better captured by the spectral coefficients. For example, the HVS is more sensitive to low-frequency coefficients, and less sensitive to high-frequency coefficients. In other words, low-frequency coefficients are perceptually significant, and alterations to those components might cause severe distortion to the original image. On the other hand, high-frequency coefficients may be considered insignificant, and processing techniques, such as compression and filtering, tend to remove high-frequency coefficients aggressively. To obtain a balance between imperceptibility and robustness, most algorithms embed watermarks in the midrange frequencies.

Watermarking algorithms were initially developed for the copyright protection of digital images and videos and subsequently they were applied to the audio data also. In the past few years, several algorithms for the embedding and extraction of watermarks in audio sequences have been presented. All these developed algorithms take advantage of the perceptual properties of the HAS in order to embed the watermark in an imperceptible manner. A broad range of embedding techniques are found in the literature which include the simple least significant bit (LSB) scheme [163], echo hiding [11], patch work algorithm [164] and spread spectrum method [87].

Following previous discussions, we see that the watermark should not be placed in perceptually insignificant regions of the media (or its spectrum), since many common signal and geometric operations affect these components. So the issue is how to insert a watermark into the perceptually significant regions of the spectrum in a fidelity preserving manner. Clearly, any spectral coefficient may be altered, provided such a modification is small. To solve this problem, the frequency domain embedding is preferred. Attacks and unintentional signal distortions are treated as noise that the immersed signal must be immune to. This is called spread spectrum watermarking [29].

In spread spectrum modulation (SSM) or CDMA communication system, one transmits a narrowband signal over a much larger bandwidth such that the signal energy present in any single frequency is undetectable. Similarly, the watermark is spread over very many frequency bins so that the energy in any one bin is very small and certainly undetectable. Nevertheless, because the watermark verification process knows the location and content of the watermark, it is possible to concentrate these weak signals into a single output with high signal-to-noise ratio (SNR). However, to destroy such a watermark one would require noise of high amplitude to be added to all frequency bins which would destroy the signal itself. Hence in SSM or CDMA based watermarking techniques [29], one embed the information by linearly combining the host media with a small pseudo-noise (PN) signal that is modulated by the watermark.

Spreading the watermark throughout the spectrum of an image ensures a large measure of security against unintentional or intentional attack: First, the location of the watermark is not obvious. Furthermore, frequency regions should be selected in a fashion that ensures severe degradation of the original data following any attack on the watermark. A watermark that is well placed in the frequency domain of an image or a sound track will be practically impossible to see or hear. This will always be the case if the energy in the watermark is sufficiently small in any single frequency bin.

7.3 Key-frame watermarking

A wide variety of image watermarking schemes may be found in literature, addressing specific application scenarios. Works on gray scale image watermarking is abundant, while the extension to the color case is usually accomplished by handling the image luminance [7, 11] or by processing each color channel separately [95].

An alternative approach to color image watermarking has been developed in [68] using which the authors embed the watermark in the *blue* channel, as the HVS is claimed to be less sensitive to changes in this band. A similar procedure is adopted in this book for the color image watermarking of the key-frames extracted from the lecture video. It is also claimed that this type of component watermarking of the color image is also robust against monochrome conversion.

Note that the IMP contains both content and non-content key-frames, as discussed in Chapter 5. In non-content key-frames, there are significant high frequency components whereas in content key-frames, it is low due to the presence of a more or less smooth background of paper or slide template on which the instructor writes. It may be noted that the imperceptibility characteristic of the watermark is critical in the case of content key-frames whereas it could be more tolerant for talking head key-frames. This is because the textual regions on the content key-frames should not deteriorate due to embedding. At the same time, any artifact on instructor's face would not affect the dissemination of knowledge through media re-creation using the IMP. So talking head frames may be watermarked with a higher modulation index. The image watermarking scheme adapted here is a DCT based one, in which the watermark bits are embedded in the midband coefficients through spread spectrum technique. The middle frequency bands are chosen because it avoids altering the visually most important parts of the image (low frequencies) and offers a better survivability against the removal through compression and noise attacks (high frequencies).

Let the blue component of the key-frame be $I_B(m,n)$. The 2-D discrete cosine transform (DCT) of an 8×8 image block $b(m,n)$, contained in $I_B(m,n)$ is defined as [43]

$$C(u,v) = \alpha(u)\alpha(v) \sum_{m=0}^{7} \sum_{n=0}^{7} b(m,n) \cos\left[\frac{(2m+1)u\pi}{16}\right] \cos\left[\frac{(2n+1)v\pi}{16}\right] \quad (7.1)$$

for $u, v = 0, 1, 2, ..., 7$, and

$$\alpha(u) = \begin{cases} \sqrt{\frac{1}{N}} & for \ u = 0 \\ \sqrt{\frac{2}{N}} & for \ u = 1, 2, ..., N-1. \end{cases} \quad (7.2)$$

Similarly, the inverse DCT is defined as

$$b(m,n) = \sum_{u=0}^{7} \sum_{v=0}^{7} \alpha(u)\alpha(v)C(u,v) \cos\left[\frac{(2m+1)u\pi}{16}\right] \cos\left[\frac{(2n+1)v\pi}{16}\right] \quad (7.3)$$

for $m, n = 0, 1, 2, ..., 7$.

The DCT allows an image to be broken down into different frequency bands, making it much easier to embed watermarking information into the middle frequency bands of an image. We define the middle-band frequencies f_M of an 8×8 DCT block as shown in Fig 7.2. f_L is used to denote the low frequency components of the block, while f_H is used to denote the high frequency components. The mid-

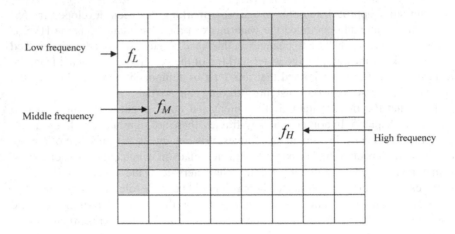

Fig. 7.2 A diagram showing different components of the DCT spectrum for an 8×8 block.

band f_M is chosen as the embedding region so as to provide an additional resistance to lossy compression techniques, while avoiding a significant modification of the cover image.

7.3.1 Embedding strategy

The block diagram of the DCT based image watermark insertion scheme is shown in Fig 7.3. In this, a 2-D watermark gets embedded in the blue channel of the image using an 8×8 block, DCT based spread spectrum technique. Let the dimensions of the original image (cover) be $M \times N$ and that of the watermark (message) be $M_w \times N_w$. The 2-D watermark is converted to a message vector before it gets embedded in the cover. There is a limit to the maximum allowable message size for the given dimensions of the cover. Since we are employing DCT based watermarking, the maximum message size should not exceed $(M \times N)/64$. To effect spread spectrum coding of the watermark, we use two separate, highly uncorrelated pseudo-noise (PN) sequences for the encoded bits 0 and 1. These PN sequences are produced by a random sequence generator with the help of a *key*.

We aim at embedding one bit from the message at each of the 8×8 blocks in the image. In doing so, the DCT of an 8×8 image block is computed first, according to equation (7.1). As already mentioned, the mid-band DCT coefficients are the best choice for watermark embedding. Hence we modulate the 22 mid-band DCT coefficients (see Fig. 7.2) by the required PN sequence. If the bit to be encoded is a 0, then *PN sequence 1* is used, otherwise *PN sequence 2* is used. A *modulation index* or *gain factor k* is used so as to control the depth of watermarking. The modulation of the DCT block is given by

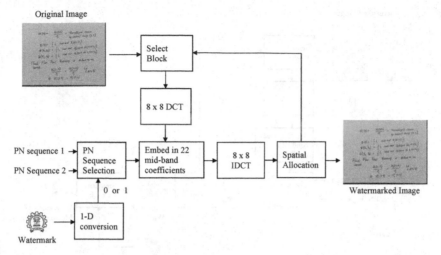

Fig. 7.3 A block diagram of the DCT based spread spectrum watermark insertion system for key-frames.

$$C(u,v) \leftarrow \begin{cases} C(u,v) + k\,r_b(u,v) & for\ u,v \in f_M \\ C(u,v) & for\ u,v \notin f_M. \end{cases} \qquad (7.4)$$

where k is the watermarking depth, $r_b(u,v)$ is the PN sequence for $b = 0$ or 1. Typical sample values of PN sequence used for zero and one, respectively, are

$$\begin{bmatrix} 1 & -1 & -1 & 0 & 0 & 0 & -1 & 0 & -1 & -1 & 1 & 1 & 0 & 0 & 0 & 0 & -1 & 1 & -1 & 1 & 0 & 0 \end{bmatrix}$$

and

$$\begin{bmatrix} 0 & 1 & 1 & 0 & 0 & 1 & 0 & -1 & 0 & 1 & -1 & 0 & 0 & 1 & 0 & 0 & 1 & -1 & 0 & 0 & 0 & 0 \end{bmatrix}.$$

Increasing k improves the robustness of the watermark at the expense of quality. Note that the DCT coefficients in the low and high frequency ranges remain unaffected. Then the watermarked DCT block is inverse transformed to yield the corresponding 8×8 image block. The process is repeated for each such image block, covering all the bits from the message to be embedded.

7.3.2 Extraction strategy

We illustrate the watermark extraction system in Fig. 7.4. In this, the algorithm looks for the presence of the watermark in the input image. As already mentioned, the blue channel of the given image is selected for decoding. Then for each 8×8 image block, the DCT is computed and the 22 mid-band coefficients are selected. The correlation of these coefficients with the two PN sequences are found out to

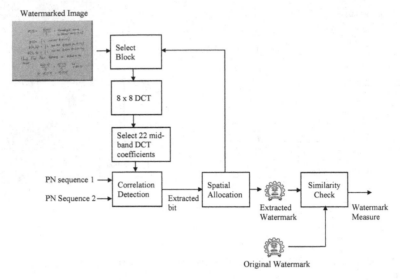

Fig. 7.4 Illustration of the DCT based spread spectrum watermark extraction system for key-frames.

decide if any of these two bits is actually there in the block of interest. This process is repeated for all the image blocks and the output bit sequence is spatially collated to form the extracted watermark. A high measure of similarity between this extracted watermark and the original one would tell whether the given image is a copyright protected one.

7.4 Audio watermarking

The speech signal of the lecturing activity is an integral part of the IMP. This also needs to be watermarked before the IMP is distributed to the users. Audio watermark also needs to sustain various attacks. According to International Federation of the Phonographic Industry (IFPI), under the constraint of imperceptibility, the watermark embedded in the audio should sustain a time stretching of $\pm 10\%$, random cropping, and common signal processing attacks like low pass filtering, requantization, etc. Condition for imperceptibility is that the signal-to-noise ratio (SNR) should be higher than 20 dB. Desynchronization attacks like time scale modification (TSM) and random cropping are the most challenging attacks in the case of audio. In TSM attacks, even under a time scaling of $\pm 10\%$, the auditory quality is still rather perfect, as the human auditory system is partly insensitive to such modifications. This makes TSM even more challenging. Generally there are two main modes of TSM [160] - *pitch invariant* and *resampling*. The pitch invariant TSM preserves audio pitch, and the resampling TSM modifies the playback speed by stretching the

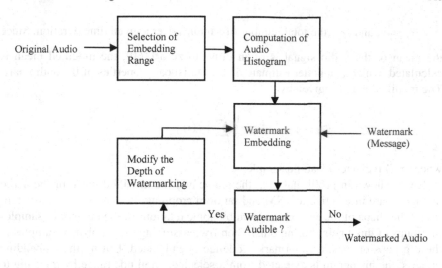

Fig. 7.5 A block schematic diagram of the histogram based audio watermark embedding framework.

audio. Robustness against random cropping is also very critical in audio watermarking. There are two types of possible cropping operations. In *local uneven cropping*, one or several parts of the audio are randomly cut, which results in a displacement of the watermark. In *global even cropping*, also known as *jittering* [71], signals are selected and cut evenly from the audio stream.

The audio signal in distance education videos is of much importance in terms of instructional content. Any attempt to crop the audio considerably will result in the loss of instructional values. Hence attacks would not, in general, involve much cropping. At the same time, high values of embedding depth for the watermarking of speech signal can be achieved since a music quality audio is actually not required for the content re-creation using IMP.

We use the histogram based audio watermarking scheme propounded in [160] the block diagram of which is given in Fig. 7.5. This method is a robust audio watermarking solution, which is claimed to be able to efficiently tackle the desynchronization attacks discussed earlier. It does so by using the insensitivity of the shape of the audio histogram and the modified mean to such attacks. The term histogram is generally used to describe the distribution of data. Audio histogram can be obtained first by temporally splitting the data and then classifying the data samples into equal-sized bins and then by counting the number of samples that fall into each bin. Hence the audio histogram may be described by

$$H = \{h(i) \mid i = 1, ..., L\} \tag{7.5}$$

where H is a vector denoting the volume-level histogram of the audio signal $F = \{f(i) \mid i = 1, ..., N\}$, $h(i)$ denotes the number of samples in the ith bin, such that

$\sum_{i=1}^{L} h(i) = N$, and N is the total number of samples over a given time duration. Since the mean of the audio signal is often close to zero [160], the modified mean is calculated, which is a better estimate of the statistical properties of the audio data. The modified mean is given by

$$\bar{A} = \frac{1}{N} \sum_{i=1}^{N} |f(i)| \qquad (7.6)$$

where $f(i)$ indicates i^{th} audio sample.

It was shown in [160] that both the shape and the modified mean of the audio histogram are insensitive to TSM and random cropping. In that work, authors represent the shape of the audio histogram by some relations of the number of samples in groups of three neighboring bins. Then by reassigning the number of samples in these groups of bins, a watermark sequence is embedded. During the embedding process, the histogram is extracted from a selected amplitude range by referring to the mean in such a way that the watermark will be able to be resistant to amplitude scaling. Consequently an exhaustive search is avoided during the watermark extraction process. For a detailed description of these watermark embedding and extraction strategies readers are advised to refer [160].

7.4.1 Multi-segment watermarking

It may be noted that in the audio watermarking algorithm based on histogram shape, if the cropped samples have a different distribution than that of the original amplitude range, the histogram shape changes significantly. The samples deleted evenly from the host audio, as is done in jittering, usually have statistically a very similar distribution, as the host audio. Thus we can say that jittering has very less effect on the histogram shape and the modified mean. In case of uneven cropping, the less the number of samples cropped, the higher is the probability of the histogram shape remaining unchanged. Thus, this algorithm, though very robust to TSM($\pm$20%), will not perform well under heavy local cropping. To tackle this problem we embed the watermark in multiple sections of the audio using the histogram shape based algorithm. By this we retain the positive aspects of the above algorithm, and use the watermarks extracted from different segments in a combined way to figure out the final watermark accurately. The final algorithm will be robust to both high degrees of TSM and local cropping attacks, and also to extensive common signal processing manipulations mentioned earlier.

We embed the same watermark into multiple segments of the host audio to handle high degrees of local cropping. The original audio is split into segments of length l seconds, and the watermark is embedded into all of these segments. These watermarked segments are joined together to get back the entire watermarked audio stream. Attacks tend to destroy the watermark embedded in each segment of the audio signal. Our goal is to extract the hidden bits from multiple such segments to reduce the detection error.

Considering only cropping attack to be present, this is expected to produce a mismatch between the actual and observed segment boundaries. If the segment boundaries in the attacked file are taken as they are (every l seconds), they will result in a high degree of error in extraction. Thus we make an effort to find the correct alignments in the received/attacked file. In order to find correct alignments, initially the search is done with a step size of l seconds and the watermark is extracted. A local cropping instance (or an absence of watermark) is indicated by a high error in detection. Then the step size is reduced and a refined search is performed, until the required detection accuracy is achieved. Now, with the obtained alignment, again the extraction is continued with a bigger step size for the subsequent segment. This is followed until a cropping instance (high degree of error) is detected.

Practically there could be other attacks also, along with random cropping, which might result in errors in the segments that are uncropped and are even found to be in correct alignments. The errors in the uncropped segments have to be less than the errors that occur at the cropping boundaries for the correct alignments to be found. To simulate this scenario and to compute the degree to which such kind of degradation can be tolerated, we have used an additive noise along with random cropping during experimentations. This is discussed in detail in the results section.

7.4.2 Decision fusion

Distortion due to random cropping helps us to find out the correct alignments of the watermarked multi-segments. As mentioned, several different types of attacks might be present which would result in errors in the segments that may be in correct alignments. Depending upon the degree of attack, multiple bits may go wrong in the correctly aligned segments. To handle this, we use *decision fusion*. It is a technique that has received considerable attention, and is well established in pattern recognition literature. By properly combining the individual outputs, we expect a better accuracy than that for each of the outputs. There are several reasons for which decisions are combined. We may have different feature sets, training sets, classification methods or training sessions, all resulting in different outputs. These outputs then can be combined, with the hope of improving the overall classification accuracy [58]. A large number of combination schemes are proposed in literature [67, 162]. A typical combination scheme consists of a set of individual classifiers and a combiner that combines their individual results to make the final decision. When an individual classifier should be invoked and how they should interact with each other is decided by the architecture of the combining scheme. Various combining schemes differ from each other in their trainability, adaptivity and requirement of the output of the individual classifiers.

Different combiners expect different types of outputs from individual classifiers [162], and the outputs may be grouped as *measurement* (or *confidence*), *rank* and *abstract*. At the confidence level, a classifier outputs a numerical value for each class indicating the belief or probability that the given input pattern belongs to that class. At the rank level, a classifier assigns a rank to each class with the highest rank being the first choice. Rank value cannot be used in isolation because the highest rank does not necessarily mean a high confidence in the classification. At the abstract level, a classifier only outputs a unique class label or several class labels (in which case, the classes are equally good). The confidence level conveys the richest information, while the abstract level contains the least amount of information about the decision being made. Some of the popular combination schemes are - maximum, minimum, voting, mean, median [67], Dempster-Shafer, and Bayes belief integrals. We have used Bayesian belief integration technique for fusing the individual decisions.

Methods of combining decisions like minimum, maximum, average and majority vote [67] do not take into account the errors that have occurred in a segment. They just consider the output labels. But in the Bayesian formulation, errors for each segment are captured by a matrix called confusion matrix [162], represented as

$$
PT_k = \begin{pmatrix} n_{11}^{(k)} & n_{12}^{(k)} & \cdots & n_{1M}^{(k)} \\ n_{21}^{(k)} & n_{22}^{(k)} & \cdots & n_{2M}^{(k)} \\ \vdots & & \ddots & \vdots \\ n_{M1}^{(k)} & n_{M2}^{(k)} & \cdots & n_{MM}^{(k)} \end{pmatrix}
$$

$k = 1, 2 \ldots K$, where each row i, corresponds to class C_i, and each column j corresponds to an event $s_k(x) = j$, meaning that in segment s_k, bit x is given a label j. Thus $n_{ij}^{(k)}$ means that $n_{ij}^{(k)}$ bits of class C_i have been assigned a label j in segment k. Thus the total number of samples/bits in a segment are

$$
N^{(k)} = \sum_{i=1}^{M} \sum_{j=1}^{M} n_{ij}^{(k)} \tag{7.7}
$$

in which the number of bits in each class C_i are,

$$
n_i^{(k)} = \sum_{j=1}^{M} n_{ij}^{(k)}, \ i = 1 \ldots M \tag{7.8}
$$

and the number of samples that are assigned a label j in s_k are

$$
n_j^{(k)} = \sum_{i=1}^{M} n_{ij}^{(k)}, \ j = 1 \ldots M. \tag{7.9}
$$

If a segment s_k has an error, then the corresponding event $s_k(x) = j$ will have certain uncertainty. With the help of the corresponding confusion matrix, such an uncertainty is expressed with a probability

$$P(x \in C_i | s_k(x) = j) = \frac{n_{ij}^{(k)}}{n_j^{(k)}}, \quad i = 1 \dots M. \tag{7.10}$$

This is also known as the belief in a proposition

$$bel(x \in C_i | s_k(x) = j) = P(x \in C_i | s_k(x) = j), \quad i = 1 \dots M. \tag{7.11}$$

With K segments, $s_1, \dots s_K$, we have K confusion matrices, $PT_1, \dots PT_K$. When we consider the same bit in all these segments, then we have K events, $s_k = j_k$, $k = 1 \dots K$. It means that in each of the segments, the same bit can have a different class label. Using the above definition, each event along with the corresponding confusion matrix will give a belief value for each class. To arrive at a single label we need to combine these individual decisions. These decisions can be fused using the Bayesian formulation.

By using a standard naïve Bayes formulation, and by assuming that events $s_1(x) = j_1, \dots s_k(x) = j_k$ are independent, an expression for belief can be derived [162]. It may be given by

$$bel(i) = \eta \prod_{k=1}^{K} P(x \in C_i | s_k(x) = j_k) \tag{7.12}$$

with η being a constant that ensures, $\sum_{i=1}^{M} bel(i) = 1$ which is calculated as

$$\frac{1}{\eta} = \sum_{i=1}^{M} \prod_{k=1}^{K} P(x \in C_i | s_k(x) = j_k). \tag{7.13}$$

In our framework, for every bit, beliefs for all the possible classes are found out using the individual segment labels and confusion matrices. Then the bit being tested is allocated to the class having the maximum belief.

7.5 Illustrative results

Different experimentations are performed to demonstrate the results of watermarking of both key-frame images and audio data which are the components of the IMP derived from lecture videos. The results of these experimentations are discussed in the following subsections.

7.5.1 Key-frame watermarking

We start with a result in which a content key-frame is watermarked with a message according to our embedding technique. Fig 7.6 shows a content key-frame, the wa-

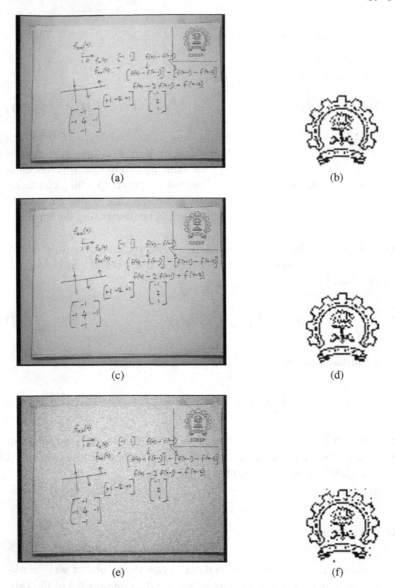

Fig. 7.6 Results of the DCT based CDMA watermarking of content key-frames. (a) is the original video frame, (b) is the original watermark, (c) is the watermarked frame (PSNR = 27.96 db), (d) is the extracted non-attacked watermark, (e) is the degraded (attacked) frame (PSNR = 22.12), and (f) is the corresponding extracted watermark (after attack).

termark, the corresponding watermarked image, the extracted watermark, the watermarked frame with some sort of attack and the corresponding extracted watermark. In this figure, the attack or degradation (PSNR of 22.12 db) is introduced by adding

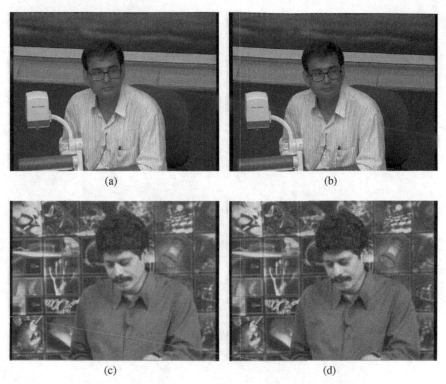

Fig. 7.7 Two talking head key-frames (a, c) and their watermarked ones (b, d). Note that the watermark is embedded quite invisibly for the watermarking depth of $k = 30$.

Gaussian noise to the watermarked image. Note that even if the watermarked color frame is degraded much by the attack, the extracted watermark is without much distortion. The algorithm was also tested for other type of attacks the results of which are given later in this section.

Now we show two original talking head key-frames along with the watermarked ones in Fig 7.7 for perceptual comparison purposes. Similarly two original writing hand key-frames along with the watermarked ones are shown in Fig 7.8. Note that the watermark is embedded sufficiently imperceptibly in these frames and so the watermarked frames (b) and (d) in both Fig 7.7 and Fig 7.8 almost resemble their original ones. The embedding depth k is fixed as 30 for all these cases.

The robustness of the method is tested against attacks like scaling and smoothing. A comparative study is done in which the performance is analyzed relative to other methods of watermarking - (a) watermark embedding by the comparison of mid-band DFT coefficients without using any CDMA technique [60] and (b) the discrete wavelet transform (DWT) based watermarking scheme [69]. The first method encodes a watermark bit in the mid-band DCT coefficients based on a comparison strategy as discussed in [60] without using any PN sequence. The second scheme

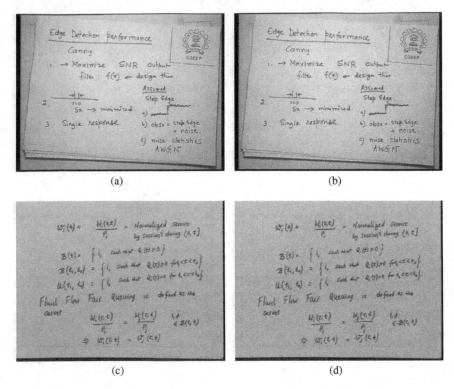

Fig. 7.8 Two writing hand key-frames (a, c) and their watermarked ones (b, d). The watermarking depth is chosen as $k = 30$ here also.

embeds watermark bits in those regions of image *details* in the DWT decomposition (horizontal detail - HL, vertical detail - LH and diagonal detail - HH) using a CDMA technique as propounded in [69]. In Fig 7.9, the robustness of the watermarking scheme for a writing hand key-frame against the above attacks are shown. A similar set of results for a talking head key-frame is given in Fig 7.10. Note that the green curve represents the suggested method while red represents the method of comparison of DCT coefficients (without using any PN sequence) and blue represents the DWT based method. From Fig 7.9 and 7.10, it could be noted that the suggested method is robust enough compared to the other two, against these attacks.

7.5.2 Audio watermarking

In order to test the robustness of the audio watermarking algorithm, random cropping along with additive white noise is used. To add white noise, Stirmark Benchmark for audio [127] is used. The strength of the noise is set to 10. Random cropping

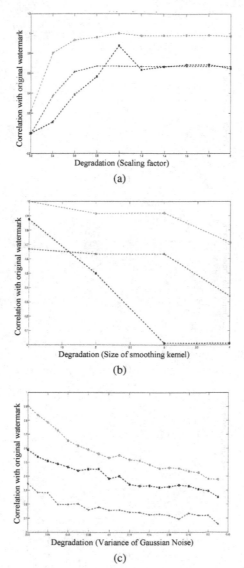

Fig. 7.9 Plots showing the performance of the watermarking scheme. (a), (b) and (c) describe the robustness of the watermarking for a writing hand key-frame against scaling, smoothing using a uniform kernel and Gaussian filtering, respectively. The green curve represents the suggested method while red represents the DCT based method without using CDMA and blue represents the DWT based method.

of the order of 20% of the segment length is applied. The analysis is carried out by varying both the length of watermark sequence and the length of audio segment. The length of watermark sequence is varied from 60 bits to 5 bits, and the audio

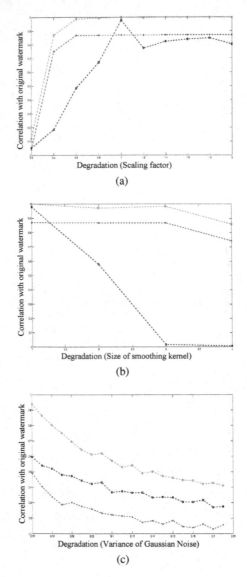

Fig. 7.10 Plots showing the performance of the watermarking scheme. (a), (b) and (c) describe the robustness of the watermarking for a talking head key-frame against scaling, smoothing using a uniform kernel and Gaussian filtering, respectively. The green curve represents the suggested method while red represents the DCT based method without using CDMA and blue represents the DWT based method.

segment length is varied from 60 seconds to 10 seconds. An audio stream from a 1 hour long video lecture is considered. This analysis is carried out in order to find

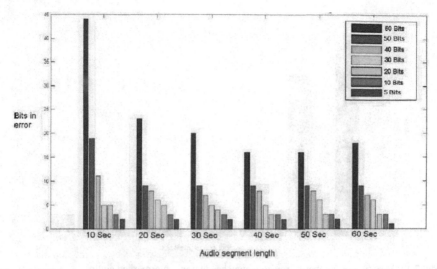

Fig. 7.11 Variation in noise tolerance with respect to the number of watermarked bits.

a useful audio segment length and watermark sequence length, which will yield a reasonable amount of robustness under the considered attacks.

The shape of the histogram calculated from a selected audio range is used to embed watermark bits. As already discussed, a group of three consecutive bins carries one watermark bit. When the length of the watermark sequence is reduced, we need less number of bins to hide them, and so the histogram is plotted with less number of bins. This results in an increase in the width of the histogram bins. When the noise dominates the host audio, some samples are modified, resulting in an amplitude change of the original sample values. In case of a larger bin width, a large range of amplitude values are crammed in a given bin. Thus even if there is an amplitude change in the noise garbled samples, the chance of them shifting to some other bin is less. Thus the shape of the histogram does not change much, resulting in a less number of bits going wrong during detection. As opposed to this when the length of the watermark is larger, the width of the histogram bins is smaller, and the same amount of noise, now makes the samples shift their bins more frequently, resulting in a change in histogram shape and a larger error during detection. Thus it is observed that noise tolerance goes on increasing as we embed less number of bits in the audio signal of a given length. This is shown in Fig. 7.11. As expected, it is seen that the maximum number of bits that can go wrong in an uncropped segment goes on reducing as the length of the watermark sequence reduces (under the same attack environment). This trend is observed at all audio segment lengths. To exploit this increasing noise tolerance, we should use a watermark of smaller length.

The number of falsely obtained alignments in the attacked watermarked audio goes on increasing as the number of watermark bits reduces. This scenario could be very helpful in the case of extraction of hidden watermark from a watermarked audio, as we get more segments, and that too with less errors, which after decision

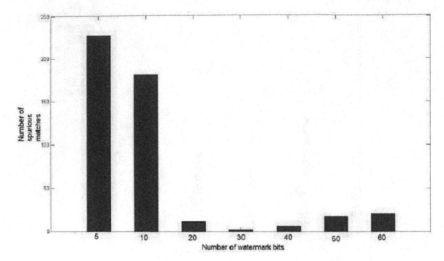

Fig. 7.12 Variation in number of spurious matches in non-watermarked audio.

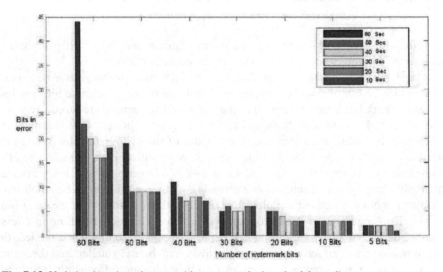

Fig. 7.13 Variation in noise tolerance with respect to the length of the audio segment.

fusion will yield the best results. But on the other hand, there is a danger of obtaining such small watermarks in just any audio files, even un-watermarked. In other words, we can say that such sequences, of very small lengths, are often not capable of being called as watermarks, as they are too frequent in any kind of host signal. Thus to get a lower limit on the length of the watermark, we analyzed the accidental occurrences of these varying length watermarks in an un-watermarked audio. The searches are performed under the corresponding noise tolerances, and we got an interesting result. For a 30-bit watermark sequence, with an error tolerance of 5 bits, we got the minimal false matches in the un-watermarked audio. This is shown

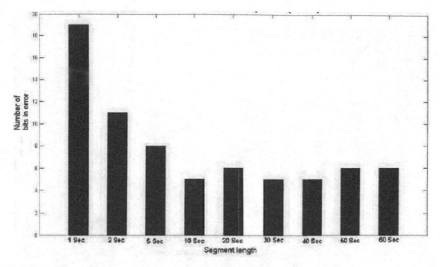

Fig. 7.14 Variation in noise tolerance for very short audio segment.

in Fig. 7.12. Thus we can say that to exploit the increasing noise tolerance, we should use watermarks of smaller lengths, but not below 30 bits.

Now we fix the length of the watermark sequence, and observe the effect of change in the length of the audio segment on the noise tolerance. In Fig. 7.13, variation in noise tolerance is plotted as the audio segment length varies. For 30 bits of watermark sequence, across different segment lengths, the variation in noise tolerance is observed to be negligible. To observe the effect in any of the length of the audio segment, we further reduced the length to 5 seconds, 2 seconds, and 1 second. Then we observed that the noise tolerance drops as the segment length is made too small, when more and more number of bits start going into error. This is shown in Fig. 7.14. So it becomes clear that this method of audio watermarking is dependent on the audio segment length for its performance.

The correctly aligned segments are then considered together and a naïve Bayes decision fusion method is used to accurately detect the watermark as described in section 7.4.2. The increasing nature of the detection accuracy as more and more segments are fused, is shown in Fig. 7.15, justifying the usefulness of the decision fusion process.

7.6 Discussions

A methodology for the copyright protection of IMP is discussed in this chapter. The key-frames and the audio track are suitably watermarked for this purpose. A block DCT based spread spectrum scheme is employed for the watermarking of key-frames. The color image watermarking is effected by marking the blue channel

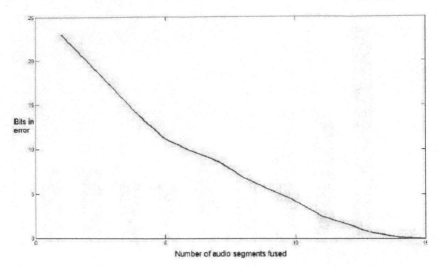

Fig. 7.15 Variation of error in number of bits with respect to the number of fused audio segments.

since the HVS is claimed to be less sensitive to changes in this band. The algorithm embeds the watermark in the midband coefficients of the 8×8 DCT block. The middle frequency bands are chosen because it avoids altering the visually most important parts of the image (low frequencies) and offers a better survivability against the watermark removal through compression and noise attacks (high frequencies). The robustness of the image watermarking scheme is tested against various kind of attacks.

The audio stream was watermarked using a multi-segment watermarking scheme which employs the audio histogram based method propounded in [160]. The decision fusion approach was used to attain a high detection accuracy. The detection accuracy was found to be high in an attack scenario consisting of 20% random cropping along with additive white noise.

The image and audio watermarking schemes effectively protect the ownership rights of the IMP. This is badly needed since the IMP as a learning material is a form of intellectual property and is to be purchased by the users of distance education. However, a number of competing IMPs dealing with similar contents but delivered by different experts and possibly owned by different service providers may exist in the database. A question naturally arises - which lecture material (IMP) should a user subscribe to? This problem is addressed in the next chapter which discusses about a novel method of creating a capsule video meant for course preview for distance education systems.

Chapter 8
Lecture Video Capsule for Course Preview

8.1 Introduction

Digital video, though fast and efficient for the dissemination of knowledge, consumes a large amount of memory and bandwidth. A large volume of information is available at various digital repositories. However, searching for the right material is a tedious and often frustrating job. This is more true for video repositories as previewing them to decide which particular video should be selected is extremely time consuming. Previewing a lecture video consumes one hour of our time. Hence making a long video short is a promising field to video researchers. Recently the Internet has experienced an increased use of digital video which opened new dimensions in the field of education and entertainment. Consequently, research and development in new multimedia technologies which will improve the accessibility of the enormous volume of stored video are inevitable [85, 114]. In the field of education, especially where the potential users are students, a fast preview of the entire lecture would be very useful for them before downloading or even buying the video for the whole course. In this context, a lecture video synopsis or *capsule* will help them for a fast preview of its content.

One may think of a lecture video synopsis as a summarized version of the original video which contains the information about the instructor and the main topics he/she deals with. These are textual information that can be easily provided to the users. But this fails to provide any information on the quality of teaching by the instructor. In our approach, we demand that a lecture video synopsis should contain the *highlights* of the entire lecture. Hence it is essentially a problem of highlight creation of the instructional activities present in the lecture video. Although there are several works reported on the highlight creation of sports video [21, 31, 119], that of an instructional video is still an ill-addressed one. In the case of sports video, interesting events always occur with an increased crowd activity by which the audio energy level is boosted much. Hence audio is an important cue for sports video highlight creation. Also, visual information of the desired highlight, e.g. scoring a goal or a boundary (home run) hit, vary much from the ambient one in sports video.

Hence almost all of the methods for sports video highlight creation use predefined models which make use of visual as well as audio signal. In [31] the authors use audio, text and visual features for the automatic extraction of highlights in sports video. The method propounded in [119] detects highlights using audio features only without relying on expensive visual computations. In [21] authors build statistical models for each type of scene shots and a hidden Markov model (HMM) is learned for each type of highlight. In [141] authors present an HMM based learning mechanism to track the video browsing behavior of users, which is used to generate fast video previews.

We observe that the scene complexity is very low for a lecture video since it consists of only a finite number of instructional activities like talking head, writing hand or slide show (electronic slides). Since an instructional activity is shot inside a classroom, its features are much different from sports or commercial video. The audio energy level is more or less same throughout the lecture and the only cue that can be used is the visual information. Moreover, domain models such as those presented in [21] and [141] cannot be employed because the highlight occurs at some point within the same activity like a talking head, writing hand or slide show, i.e., the highlights are not climax based. As such, there is no such concept of a climax in instructional videos. Hence a new strategy needs to be developed to define the highlights in a lecture video. We base our approach on following observations. In case of non-content segments like talking head, a perceptual visual quality [131, 154] alone should suffice to provide a sense of look and feel of the instructor to the viewers. In case of content segments like writing hand or slide show, a quality measure based on both the visual quality and pedagogic content is to be used to extract the instructional highlights. Hence for defining the quality of the content frame, we suggest a method which utilizes statistical features of both the textual contents and their lay out in a given frame. The textual features measure the clarity (legibility) of the textual regions and the lay out features measure how well the frame is filled across the entire document page, helping us to identify frames with high pedagogic content. Hence we select those video snippets in which the instructor describes about the contents of a well written hand made or electronic slide as the highlights of content segments. The high quality clips from both the content and non-content segments are used in suitable proportions to create the video highlights. Audio coherence is maintained while selecting these clips during media re-creation.

We start with a brief description of the method of creating the lecture video capsule. The methods of clip selection from content and non-content segments are separately treated in subsequent sections. Then towards the end of this chapter, we explain the results of the quality metric for visual quality assessment of both content and non-content frames and then summarize the work with few discussions.

The objective is to create the highlights of a lecture, which essentially contain the pedagogic highlights of the original instructional video. Accordingly we have built a complete system for this purpose, the block diagram of which is given in Fig. 8.1. It works with the following distinct steps for the creation of highlights of instructional video:

(a) Shot detection and recognition,

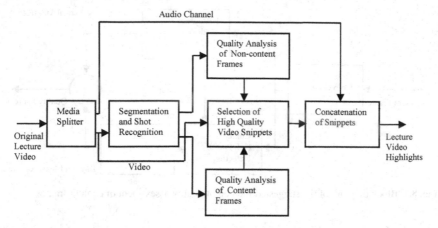

Fig. 8.1 Block diagram of the developed system of creating the highlights in an instructional video.

(b) Clip selection from content segments,
(c) Clip selection from non-content segments, and
(d) Capsule preparation.

Since the shot detection and recognition methods are already dealt with in Chapter 4, here we start with the clip selection from content segments. Note that the non-content frames provide a sense of situational awareness to the viewers and so are only of psychological importance in enhancing the dissemination capacity of the instructional video.

8.2 Clip selection from content segments

Content frames which are crucial in instructional video are very rich with text-filled regions. This fact should be taken into account while selecting representative video clips from content segments. The method proposed in [154] and discussed in section 5.5 of chapter 5 performs poorly on images with textual contents. Since our objective is to select those content clips which are well written, subject to a predefined proportion of mixing, the HPP based method given in Section 5.2.2 of Chapter 5 is also not suitable for content quality assessment here. The reason for this is given towards the end of this section after we explain the methodology of content clip section. To make the picture more clear, see the comparison of lecture video capsule (LVC) with instructional media package (IMP) which is already given in Section 2.2 of Chapter 2.

Here we adopt a different strategy for selecting relevant content clips. It is based on the *quality* of the content frames. In order to define the quality of such frames, one should consider both the clarity of text and the amount of writing (pedagogic content) on the entire frame. We use the intensity histogram to derive statistical

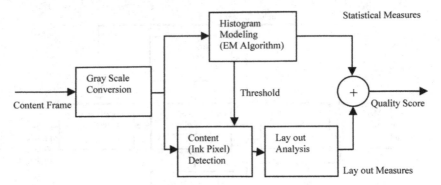

Fig. 8.2 Block diagram of the suggested method of quality assessment of content frames.

measures of clarity of writing and the horizontal projection profile (HPP) [70, 137] of the ink pixels [84] to yield layout based measures of pedagogic content. A combination of both is used to define the quality score of content frames. Hence we suggest a new method in which four separate scores derived from these features are used for the quality assessment of content frames. A block diagram of the suggested algorithm is shown in Fig. 8.2. In the figure the content frame is first converted to gray scale and then subjected to histogram modeling. The statistical measures for quality is obtained at this stage along with the threshold for ink pixel detection of the content frame. Then a layout based analysis of the ink pixel detected frame yields another pair of measures for quality. Finally the statistical measures are combined with the layout based measures to yield a quality score for the content frame which is used for the clip selection. The sub-sampling of frames as explained in section 5.2 is employed here also to reduce computation.

8.2.1 Histogram-based measures

Since color plays no role in defining the content, the frame is first converted into gray scale and the histogram $h(i)$ is computed. As already mentioned in section 5.2.1, it is modeled by a bimodal Gaussian mixture [44], with probability density function (PDF)

$$h(i) = \frac{\varepsilon_1}{\sqrt{2\pi}\,\sigma_1} e^{-\frac{1}{2}\left(\frac{i-\mu_1}{\sigma_1}\right)^2} + \frac{1-\varepsilon_1}{\sqrt{2\pi}\,\sigma_2} e^{-\frac{1}{2}\left(\frac{i-\mu_2}{\sigma_2}\right)^2} \tag{8.1}$$

where i is the intensity level, ε_1 defines the proportion of the mixture components, μ_1 is the foreground mean, μ_2 is the background mean, σ_1^2 is the foreground variance and σ_2^2 is the background variance. Let the parameter vector be $\theta = (\mu_1, \mu_2, \sigma_1, \sigma_2, \varepsilon_1)$. We have to estimate θ first so as to compute the histogram based measures of content quality and also to find the threshold T required for the layout based analysis. The threshold T for detecting the ink pixels was earlier (in Section 5.2.1) computed using the Kullback-Leibler (KL) divergence measure. The same method may be used to estimate the model parameters θ also. However, due to the truncation of the component distribution explained in Section 5.2.1, the quality of the estimates of θ is not that accurate. Hence we adopt an expectation-maximization (EM) based method [32, 90] to estimate the model parameters. We start with the generalized equations

$$h(i) = \sum_{j=1}^{2} \varepsilon_j p(i|j), \tag{8.2}$$

where

$$p(i|j) = \frac{1}{\sqrt{2\pi}\,\sigma_j} e^{-\frac{1}{2}\left(\frac{i-\mu_j}{\sigma_j}\right)^2}, \quad j = 1, 2. \tag{8.3}$$

and

$$\sum_{j=1}^{2} \varepsilon_j = 1. \tag{8.4}$$

Hence we have $h(i) \rightarrow h(i|\theta)$ and the estimation problem aims at finding the optimum vector $\theta^* = (\mu_1^*, \mu_2^*, \sigma_1^*, \sigma_2^*, \varepsilon_1^*)$ that maximizes the likelihood function

$$\theta^* = arg\,\max_{\theta} L(\theta),$$

where

$$L(\theta) = h(i|\theta),$$

and the pixel intensity values are treated as an *iid* samples of a random variable.

The EM algorithm aims at the maximum likelihood estimation of θ from the observed variables but without the exact knowledge about their probability density function. For this, it computes the joint density of these variables and some hidden variables. In the case of Gaussian mixture, the hidden variables are the kernels, the input samples statistically belong to, while each EM step provides an improved estimate of the parameters μ_j's, σ_j's and ε_j of each kernel j ($j = 2$). These iterative formula can be shown [115] to be

$$\varepsilon_j^{(t+1)} = \frac{1}{256} \sum_{i=0}^{255} P(j|i) \tag{8.5}$$

$$\mu_j^{(t+1)} = \frac{\sum_{i=0}^{255} P(j|i)\,i}{\sum_{i=0}^{255} P(j|i)} \tag{8.6}$$

$$\sigma_j^{2\,(t+1)} = \frac{\sum_{i=0}^{255} P(j|i)\,(i - \mu_j^{(t+1)})^2}{\sum_{i=0}^{255} P(j|i)} \tag{8.7}$$

where

$$P(j|i) = \frac{p(i|j)\, p(j)}{p(i)} = \frac{\varepsilon_j^{(t)}\, p(i|j; \mu_j^{(t)}, \sigma_j^{(t)})}{\sum_{k=1}^{2} \varepsilon_k^{(t)}\, p(i|k; \mu_k^{(t)}, \sigma_k^{(t)})}. \tag{8.8}$$

The advantage of using the EM framework is that it is easily programmable and satisfies a monotonic convergence property. In case of Gaussian mixtures well separated in component means, the EM algorithm performs quite well [18, 86]. Once the parameters μ_1, σ_1, μ_2, σ_2 and ε_1 are computed, the optimum threshold T for ink pixel detection is to be found out. For a given threshold T, the overall probability of error is

$$e(T) = (1 - \varepsilon_1) \sum_{i=0}^{T-1} p(i|j = 2) + \varepsilon_1 \sum_{i=T}^{255} p(i|j = 1). \tag{8.9}$$

The value of T for which $e(T)$ is minimum is given by solving [44] the equation

$$\beta_2 T^2 + \beta_1 T + \beta_0 = 0 \tag{8.10}$$

where

$$\beta_2 = \sigma_1^2 - \sigma_2^2$$

$$\beta_1 = 2(\mu_1 \sigma_2^2 - \mu_2 \sigma_1^2)$$

and

$$\beta_0 = \mu_2^2 \sigma_1^2 - \mu_1^2 \sigma_2^2 + 2\sigma_1^2 \sigma_2^2 \ln(\sigma_2 \varepsilon_1 / \sigma_1 (1 - \varepsilon_1)).$$

The threshold T is used for extracting the ink pixels [84] in the content frame prior to the layout analysis. By the process of ink pixel detection, the pixels in the content frames are converted into *ink* and *paper* pixels, corresponding to 0 and 255 values of gray levels, respectively, according to the equation (5.1). Sample frames of handwritten slides from different videos are shown in Fig. 8.3 and the corresponding values of the statistical parameters are given in Table 8.1. It may be seen that ink pixels occupy less than 10% of all pixels in a content frame. Further, we note that even though there is not much variation in the illumination, there is still a substantial change in the value of threshold T.

The mean and variance of the foreground and background can be used to define the clarity of content frame. Hence we define the following terms which contribute to defining the quality:

$$\text{Mean Separation } (C) = \frac{|\mu_2 - \mu_1|}{255}. \tag{8.11}$$

$$\text{Average Sharpness } (S) = \frac{2}{\sigma_1 + \sigma_2}. \tag{8.12}$$

If the instructor writes with a good contrast and sharpness on the paper or board, the corresponding video frames will be better legible (this is also dependent on the illumination of the object on video capture) and the measures obtained by equations (8.11) and (8.12) will be high for such frames. On the other hand, if the slides are prepared through poor and smeared markings and/or are captured under bad

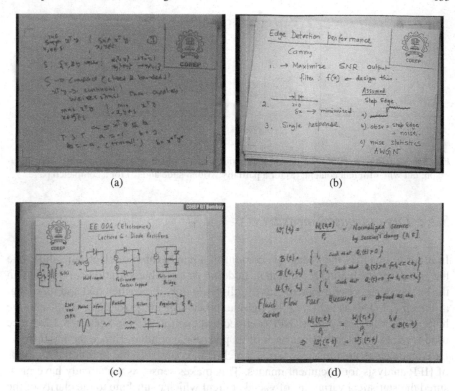

(a)

(b)

(c)

(d)

Fig. 8.3 Examples of handwritten content frames from different lecture videos. Note that the percentage of ink pixels is very low in these frames.

Table 8.1 Typical values of model parameters for the content frames shown in Fig. 8.3.

parameter	μ_1	σ_1	μ_2	σ_2	ε_1	T
frame (a)	89.41	22.50	141.03	3.58	0.09	121
frame (b)	102.71	30.90	207.77	21.60	0.07	148
frame (c)	115.90	21.65	193.01	12.94	0.12	151
frame (d)	105.32	24.55	184.43	5.90	0.07	145

illumination conditions, the corresponding video frames may not be legible and the above measures will be very low for them. Hence mean separation and average sharpness directly contribute to defining the quality of the content frames. C and S take care of the legibility aspect only and in order to include the spatial arrangement of the content, we now use the lay out based analysis to derive another pair of features.

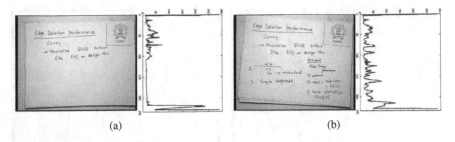

(a)   (b)

Fig. 8.4 Variations in HPP for a pair of handwritten slides which belong to the same video paragraph, but with different content: (a) for a partially written slide and (b) for a completed page.

8.2.2 Lay out-based measures

In order to quantify the pedagogic content of a handwritten or electronic slide, one has to extract the textual regions in it. In doing so, possible variations of intensity in the background are eliminated, thus highlighting only the written text and its spatial arrangement. This is achieved through the process of ink pixel detection. Then the horizontal projection profile (HPP) defined by equation (5.7) in Section 5.2.2 can be used to yield quantitative measures of its pedagogic content. Hence the input to the HPP analyzer is the binarized frame which is in contrast to the traditional methods of HPP analysis for document images. This makes sense, as we already have measured the statistical variations of visual content which contribute to the clarity of the content frame. As the instructor goes on writing on the page, the corresponding HPP also grows. Diagrams illustrating the nature of HPP variations of handwritten slides are shown in Fig. 8.4. Comparing Fig 8.4 (a) and (b), one may note that the HPP variations are negligible over those lines in which there appears no written text.

It may be noted that the HPP of an ink pixel detected frame is denoted by $F(m)$, which is an array of M elements. Since the construction of HPP is through the summing up of ink pixels along the horizontal direction, the horizontal location of ink pixels are lost. Hence the HPP of a slide which is written only through a few columns on the left side looks like that of the HPP of a slide for which the same text appears on the right side or the middle portion of the slide. In order to capture the extent to which the frame is filled *uniformly* by ink pixels across the entire document, we propose a partitioned HPP analysis. That is, we do not consider the HPP of the frame as a whole, but use those of the spatial partitions of the frame to improve the performance [27]. We observe through experimentation that a choice of 8×4 blocks to partition the image frame is quite ideal for such a partitioned HPP analysis, in which a block has an average size of 40×120 pixels that can be treated as an elementary *pedagogic block* (PB) in the content frame. Thus the ink pixel detected frame is partitioned into 8×4 equal and non-overlapping blocks. For each block, HPPs are constructed as shown in Fig. 8.5. Also, we normalize the elements of the HPP arrays in [0,1] such that uniformity is preserved for the quality scores for frames from

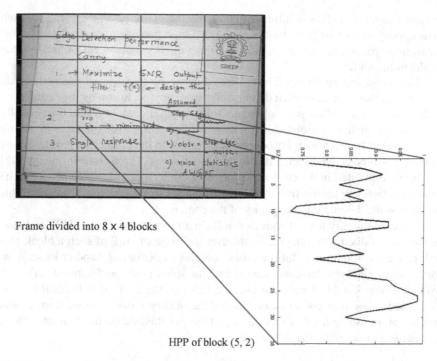

Frame divided into 8 x 4 blocks

HPP of block (5, 2)

Fig. 8.5 A content frame partitioned into 8×4 equal and non-overlapping blocks and one example of the HPP of a block.

different videos. Otherwise, the layout measures will also depend on the size $M \times N$ of the video frame.

We extract two features from these partitioned HPPs which measure the content and its organizational goodness. The first one is obtained by calculating the energies of the HPPs of the blocks by

$$E_b = \frac{1}{K} \sum_{m=1}^{K} |F_b(m) - \hat{F}_b|^2, \quad b = 1, 2, ..., 32. \tag{8.13}$$

where E_b is the energy of the HPP of the block b, $F_b(m)$ is the HPP of the block b and $\hat{F}_b$ is the average value of HPP of block b and K is the vertical size of of a block. A lower value of energy indicates predominantly black or white patches while a high value indicates well-written block. Now the energies of individual blocks are added to get the overall energy E of a frame which is a measure of content. It is defined as

$$\textit{Pedagogic Content } (E) = \sum_{b=1}^{8 \times 4} E_b. \tag{8.14}$$

If a content frame contains only a few lines of texts which occupy only a few blocks, the resulting total energy or the pedagogic content of the frame will be very low. On the other hand, if a slide is fully written across the entire page, the corresponding texts occupy almost all 8×4 blocks, yielding a high pedagogic content. Hence this measure of pedagogic content is crucial for the quality assessment of handwritten or electronic slides.

The second feature obtained from HPPs aims at the selection of relatively *clean* frames, i.e., those frames which do not suffer from any occlusion by the instructor's body parts or a dangling pen or pencil. If the original content frame occasionally contains any such occlusions, this results in corresponding black patches in the ink pixel detected frame, which introduces large values in the projection profile. A few of the 8×4 blocks get affected due to such occlusions. The picture-in-picture [48] facility incorporated in the content segments of the lecture video also results in the same situation. A quantitative measure is required to be introduced to solve this problem while calculating the quality of the content.

A block with only textual regions results in a normal variation of HPP as found in the case of document images. So the average value of HPP of such a block also will be a comparable one. But if a block contains an occluding hand or its shadow, the resulting HPP will have an unusual amount of ink pixels and hence a very high average value of HPP. Hence the average value of the HPP of individual blocks, $\hat{F}_b$ is used as another cue to de-emphasize the quality score of those frames with patches of occluding hand or picture-in-picture. For this, we perform an inter-block comparison to calculate the difference

$$d = \max_b \hat{F}_b - \min_b \hat{F}_b.$$

If d is small, all blocks are quite clean and free from patches. If d is high, there is a high likelihood of the presence of a patch or occluding objects. Hence d is an inverse measure of content quality. We define another measure which contributes to the cleanliness of the content as

$$Content\ Cleanliness\ (G) = 1/d. \tag{8.15}$$

The above term is very relevant in determining the overall quality score of content frames, since it is responsible for removing occlusions, by which an improved dissemination of the written information to the viewers is achieved. Pedagogic content and content cleanliness are layout based measures which are also the constituent terms of the quality score of content frames.

8.2.3 Quality assessment

We have defined four terms - mean separation (C), average sharpness (S), pedagogic content (E) and content cleanliness (G) - for the quality assessment of content frames, each capturing different aspects of the quality of a content frame. Now a weighted sum of these individual scores are taken to get the final quality score

$$Q_C = \alpha_1 E + \alpha_2 S + \alpha_3 C + \alpha_4 G. \tag{8.16}$$

These weights α_i s are estimated using subjective test data of content frames. This is done using Roccio algorithm [124, 88] which generally finds application in relevance feedback [121] for content based image retrieval (CBIR).

Relevance feedback (RF) is a technique that takes advantage of human-computer interaction to refine high level queries represented by low level features. It is used in traditional document and image retrieval for automatically adjusting the relevance of the extracted features to a query image using information feedback from the user. In the application of image retrieval, the user selects relevant images from previously retrieved results and provides a preference weight for each relevant image. The weights for the low-level feature, i.e., color and texture, etc., are adaptively updated based on the user's feedback. In our work, E, S, C, and G are the features whose weights α_i's are to be appropriately tuned. The user is no longer required to specify a precise weight for each feature while calculating the quality score Q_C. Taking $E = Q_1$, $S = Q_2$, $C = Q_3$ and $G = Q_4$, the overall quality measure of an estimated frame I as defined by equation (8.16) is expressed as

$$Q_C(I) = \sum_{i=1}^{4} \alpha_i Q_i(I) \tag{8.17}$$

where α_i is the low-level feature weight and $Q_i(I)$ is the quality term associated with an individual feature. Let $A = [\alpha_1, \alpha_2, \alpha_3, \alpha_4]$ be the weight vector and $Q_{Cj} = [Q_{1j}, Q_{2j}, Q_{3j}, Q_{4j}]$ be the quality terms for a frame I_j. If the sets of quality terms for high quality frames (I_H) and those for poor quality frames (I_P) are known, the optimal weights can be shown to be [125, 120]

$$A_{OPT} = \frac{1}{N_H} \sum_{j \in I_H} Q_{Cj} - \frac{1}{N_T - N_H} \sum_{j \in I_P} Q_{Cj} \tag{8.18}$$

where N_H is the number of high quality frames and N_T is the total number of frames. I_H and I_P will not be known in advance, practically. Hence the user interface allows a relevance feedback to approximate I_H and I_P to I'_H and I'_P, respectively. Here quality score is computed for all frames in the video and the user marks subsets of these frames to be subjectively of high quality and poor quality. Then the original weight vector A can be modified by emphasizing the relevant terms and de-emphasizing the non-relevant terms as

$$A' = r_1 A + r_2 \left(\frac{1}{N'_H} \sum_{j \in I'_H} Q_{Cj} \right) - r_3 \left(\frac{1}{N'_P} \sum_{j \in I'_P} Q_{Cj} \right) \tag{8.19}$$

where r_1, r_2 and r_3 are empirically chosen Rocchio coefficients [124, 88] which determine the speed at which the current weight vector A moves towards the relevant class and away from non-relevant class, N'_H is the number of observed high quality frames and N'_P is that of poor quality frames. Every entity is initially of same impor-

Table 8.2 Comparison of the results of objective and subjective tests for content quality for the content frames shown in Fig. 8.3.

	objective quality score	mean opinion score (MOS)
frame (a)	4.3286	1.5294
frame (b)	8.0264	4.8823
frame (c)	7.4385	3.9411
frame (d)	5.2109	2.9411

tance, i.e., we initialize α_i's as a set of no-bias weights of 1/4. A' approaches A_{OPT} as the relevance feedback iteration continues.

Using these updated weights obtained from a training lecture video, the quality assessment for all other content frames are done. The frame positions with high quality scores are noted for the selection of video clips for highlight creation. As explained, *well-written* content frames are rated with high quality scores. The variation in score will be mainly due to the variation in the ink pixels present since the statistical parameters do not show much variation among the frames within a video paragraph. But if our metric is applied to content frames from different videos, there is significant contribution from all the constituent terms towards the quality score. A comparison of our metric with the mean opinion score (MOS) from subjective tests for the content frames shown in Fig. 8.3 are given in Table 8.2.

One has to note that the method explained above is totally different from the procedure given in Section 5.2 of Chapter 5 for the key-frame selection to produce the IMP. The procedure described in Chapter 5 uses a content-dissimilarity based candidate key-frame selection followed by a similarity based redundancy elimination. But here for the content clip extraction we use a method which blindly selects well-written frames based on quality scores of the content frames without considering the dissimilarity in content between two frames. If this method is adopted for key-frame selection, there is a possibility of missing some distinct less-content key-frames, which would create problem on media re-creation. So this method is not capable of assuring whether the contents in the selected frames are dissimilar or poorly written frames are neglected irrespective of their distinctness. This is not a major problem when extracting the clips for creating a preview capsule. But as far as key-frame selection is concerned, the algorithm should select all the distinct content key-frames from the original video. Therefore there we go for a dissimilarity based key-frame selection using the HPP of ink pixels, as different from the method explained above. That method is able to select all the distinct key-frames from the content segments of the video. Besides, the clarity of written content could be ignored for key-frame selection as we define the key-frames as the slide which is complete in writing and minimally occluded by the instructor. Hence only the content aspect needs to be checked for the selection of representative frames. Queuing the suspected key-frames followed by redundancy elimination is a more feasible method for key-frame selection in content segments of the video. That is why we employed the HPP of the whole frame for content discrimination in Chapter 5,

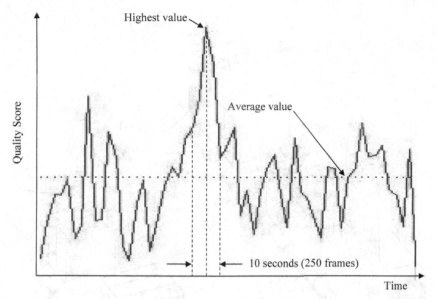

Fig. 8.6 An illustration showing the selection of a video snippet from the local quality scores of frames.

whereas in this chapter, we perform a partitioned HPP analysis to check only the completeness of the slide.

8.3 Clip selection from non-content segments

Our objective is to represent a talking head segment by the best quality video snippet selected from that segment during capsule preparation. This is done using the no-reference perceptual quality assessment method proposed in [154] which has already been explained in Section 5.5 of Chapter 5. Our assumption is that for non-content segments like talking head, mere visual quality will suffice to define *highlights*. Hence the quality score defined by equation (5.19) is computed for the frames from the talking head segment of the instructional video and the frame positions with locally high quality scores are noted for the selection of video clips. Also, to reduce computation, we sub-sample the frames as mentioned in Section 5.2.

8.4 Capsule preparation

The locations for high quality frames yield the temporal markings around which the desired video clips are to be selected, along with the audio. There are two points to

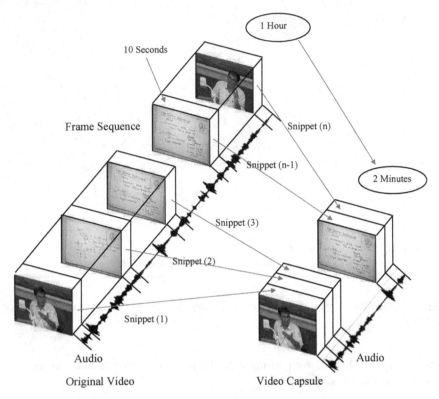

Fig. 8.7 A schematic diagram showing the sequence of the original video and the re-created media.

be taken care of during clip selection. First one is the temporal duration of a clip. It should be neither too short such that it is incapable of conveying any pedagogic content, nor too long such that the variety would be lacking in the generated video capsule. The second one is the proportion in which the snippets from the different classes of instructional activities are to be combined to produce the synopsis.

On media re-creation, we select those frames corresponding to ±5 seconds around these instants of high quality frames provided there is no shot change within this period, to produce the highlight. This is illustrated in Fig. 8.6. The choice of a 10 sec long window is meant to convey the instructional content to the viewer for a single highlight. We select such 10 sec windows around each of the prominent local peaks in the quality measure. On subjective evaluation, we found that an appropriate temporal proportion of the recognized classes during synopsis creation is 1:3 if there are only two classes, namely talking head and writing hand or slide show. If talking head, writing hand and slide show classes all occur, then an appropriate ratio was found to be 1:2:2.

As a matter of fact, the first clip shown is always the talking head to show the instructor, followed by the clips from the various segments of the video in the proportions already mentioned. The audio stream is kept in perfect synchronism with

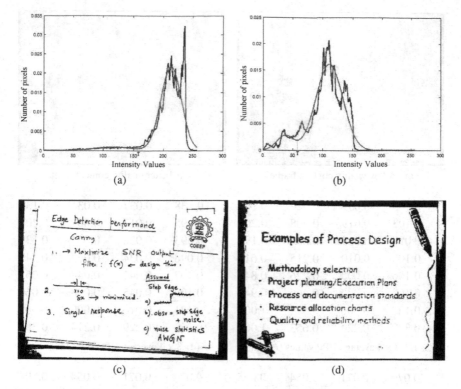

Fig. 8.8 Results of histogram modeling and ink pixel detection. (a) & (b) show the observed and estimated histograms for Fig. 8.4(b) and Fig. 8.4(c), respectively and (c) & (d) are the corresponding ink pixel detected frames. $T = 156$ in (a) and $T = 60$ in (b).

the visual data. Typically a 1 hour lecture may contain 10 to 12 video clips, each of about 10 second duration, which yields a lecture video synopsis of approximately 2 minutes as shown in Fig. 8.7.

8.5 Illustrative results

We have experimented extensively on lecture videos of different instructors, each lasting about 55-60 minutes. The frame rate for the video is 25 frames per second. The content portions in some of the videos comprise of both printed and hand written slides. Some of the videos consist only of hand written pages while the rest contains only slide shows (electronic slides).

In the case of content frames, the quality assessment process starts with the bimodal GMM assumption of the histogram of the frame. The EM algorithm iteratively computes foreground and background means and variances along with the proportion of the mixture components. From these data, an optimum threshold T

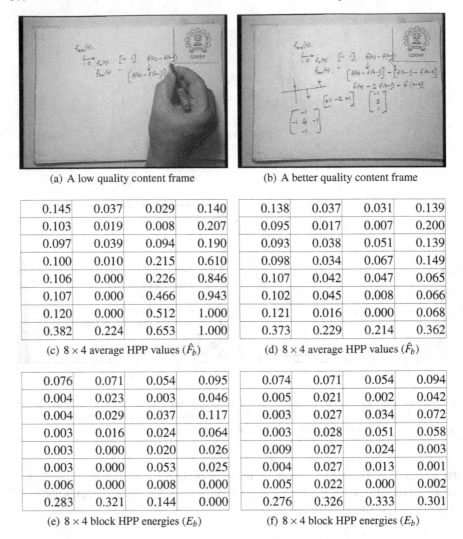

(a) A low quality content frame   (b) A better quality content frame

0.145	0.037	0.029	0.140
0.103	0.019	0.008	0.207
0.097	0.039	0.094	0.190
0.100	0.010	0.215	0.610
0.106	0.000	0.226	0.846
0.107	0.000	0.466	0.943
0.120	0.000	0.512	1.000
0.382	0.224	0.653	1.000

(c) 8×4 average HPP values $(\hat{F}_b)$

0.138	0.037	0.031	0.139
0.095	0.017	0.007	0.200
0.093	0.038	0.051	0.139
0.098	0.034	0.067	0.149
0.107	0.042	0.047	0.065
0.102	0.045	0.008	0.066
0.121	0.016	0.000	0.068
0.373	0.229	0.214	0.362

(d) 8×4 average HPP values $(\hat{F}_b)$

0.076	0.071	0.054	0.095
0.004	0.023	0.003	0.046
0.004	0.029	0.037	0.117
0.003	0.016	0.024	0.064
0.003	0.000	0.020	0.026
0.003	0.000	0.053	0.025
0.006	0.000	0.008	0.000
0.283	0.321	0.144	0.000

(e) 8×4 block HPP energies (E_b)

0.074	0.071	0.054	0.094
0.005	0.021	0.002	0.042
0.003	0.027	0.034	0.072
0.003	0.028	0.051	0.058
0.009	0.027	0.024	0.003
0.004	0.027	0.013	0.001
0.005	0.022	0.000	0.002
0.276	0.326	0.333	0.301

(f) 8×4 block HPP energies (E_b)

Fig. 8.9 Results of block based HPP analysis for handwritten content frames. (a), (c) & (e) are one set of results for a low quality content frame and (b), (d) & (f) form another set for a better quality content frame.

for ink pixel detection is found out. A typical estimate of the statistical parameters is given by $\mu_1 = 79.56$, $\mu_2 = 150.63$, $\sigma_1 = 33.83$, $\sigma_2 = 5.30$, $\varepsilon = 0.21$ and the corresponding T is 132. The threshold values show a range from 60 to 160, depending on the clarity of writing and luminance conditions. The model parameters $\{\mu_1, \mu_2, \sigma_1, \sigma_2, \varepsilon\}$ are used to compute the mean separation C and average sharpness S which show a typical range of values between 0.2 to 0.8 and between 0.02 to 0.06, respectively. Some sample results showing the observed and estimated his-

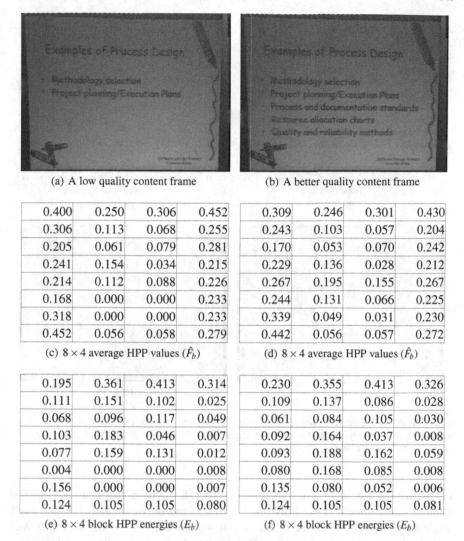

(a) A low quality content frame   (b) A better quality content frame

0.400	0.250	0.306	0.452
0.306	0.113	0.068	0.255
0.205	0.061	0.079	0.281
0.241	0.154	0.034	0.215
0.214	0.112	0.088	0.226
0.168	0.000	0.000	0.233
0.318	0.000	0.000	0.233
0.452	0.056	0.058	0.279

(c) 8×4 average HPP values ($\hat{F}_b$)

0.309	0.246	0.301	0.430
0.243	0.103	0.057	0.204
0.170	0.053	0.070	0.242
0.229	0.136	0.028	0.212
0.267	0.195	0.155	0.267
0.244	0.131	0.066	0.225
0.339	0.049	0.031	0.230
0.442	0.056	0.057	0.272

(d) 8×4 average HPP values ($\hat{F}_b$)

0.195	0.361	0.413	0.314
0.111	0.151	0.102	0.025
0.068	0.096	0.117	0.049
0.103	0.183	0.046	0.007
0.077	0.159	0.131	0.012
0.004	0.000	0.000	0.008
0.156	0.000	0.000	0.007
0.124	0.105	0.105	0.080

(e) 8×4 block HPP energies (E_b)

0.230	0.355	0.413	0.326
0.109	0.137	0.086	0.028
0.061	0.084	0.105	0.030
0.092	0.164	0.037	0.008
0.093	0.188	0.162	0.059
0.080	0.168	0.085	0.008
0.135	0.080	0.052	0.006
0.124	0.105	0.105	0.081

(f) 8×4 block HPP energies (E_b)

Fig. 8.10 Results of block based HPP analysis for slide show frames. (a), (c) & (e) are one set of results for a low quality content frame and (b), (d) & (f) form another set for a better quality content frame.

tograms for content frames along with the threshold and the resulting ink pixel detected frames are given in Fig. 8.8. Also, the pseudo-ink pixels due to the presence of unwanted shades/black patches at the spatial borders of video frames which do not contribute to the semantic content could be eliminated before further computations are made on ink pixel detected frame, as explained in section 5.7.

The 8×4 block based HPP analysis is found to be effective in quantifying the spatial distribution of contents and hence the quality of a frame. Two sets of results

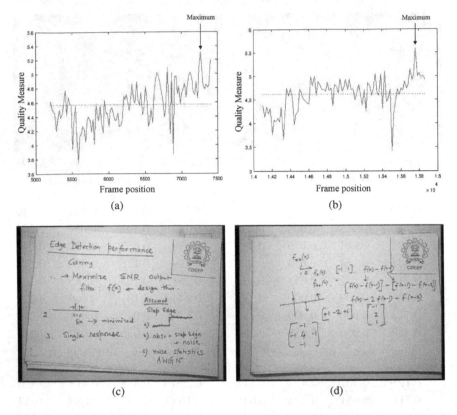

Fig. 8.11 Results of no-reference quality assessment of hand written content frames. (a) and (b) are the plots of quality scores for different segments of the same video and (c) and (d) are the corresponding locally best quality frames.

of this for hand written slides are shown in Fig. 8.9. Fig. 8.9(a) is a poor quality content frame since it is not fully filled with text and is also occluded by the hand while Fig. 8.9(b) is higher in quality, since it is more filled with contents. Fig. 8.9(c) and (d) give the average values of the HPPs of the blocks and Fig. 8.9(e) and (f) show the corresponding energies of the HPPs of blocks. Note that a white patch (block) in the frame has an average HPP value ($\hat{F}_b$) of 0 and has a resultant HPP energy (E_b) of 0 since it is devoid of ink pixels. Hence the frame in Fig. 8.9(a) results in plenty of such values (zero energies) which signify a low pedagogic content E. Also, an occluding hand in Fig. 8.9(a) yields a few blocks with very high or saturated average values of HPP in Fig. 8.9(c) by which its quality score is reduced much by the content cleanliness term G. Similar results for slide show content frames from another video are shown in Fig. 8.10.

The statistical parameters derived from the histogram measure the clarity of the textual regions with respect to their local background and the layout based parameters measure the spatial distribution of contents across the entire frame. This quality

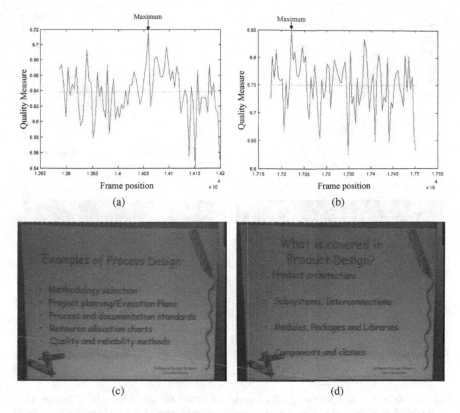

Fig. 8.12 Results of no-reference quality assessment of slide show frames. (a) and (b) are the plots of quality scores for different segments of the same video and (c) and (d) are the corresponding locally best quality frames. Note that the variation in quality score is much smaller compared to that in the case of writing hand frames.

assessment of content frames based on the four component feature model is performed with the following weights: $\alpha_1 = 1$, $\alpha_2 = 25$, $\alpha_3 = 4$, $\alpha_4 = 0.4$, which have been obtained through Roccio algorithm. The results of the quality assessment of handwritten content frames are shown in Fig. 8.11 and those of slide show content frames are shown in Fig. 8.12. From Fig. 8.11, it may be noted that frames 7286 and 15782 offer locally highest quality scores for the handwritten slides and in Fig. 8.12, frames 14062 and 17224 do the same for electronic slides. Referring to Fig. 8.9 again, we can see the increase of quality score with respect to the increase in textual content. The pair of frames shown in Fig. 8.9 (a) and (b) are taken from the same video segment but at different frame instances 3460 and 10800, respectively. It is observed that the quality score Q_C defined by equation (8.16) is found to be increased from 3.28 to 7.12 as we go from frame (a) to (b). Similarly for the frames shown in Fig. 8.10 (a) and (b), the quality score is found to be increased from 5.21 to 8.13 as (a) and (b) represent the frame instances 12660 and 14130, respectively,

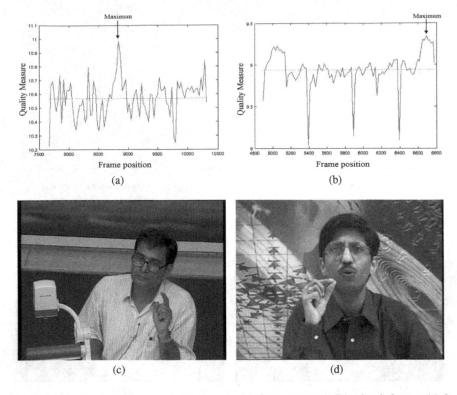

Fig. 8.13 Results of no-reference quality assessment of non-content (talking head) frames. (a) & (b) are the plots of the quality measure for different frame sequences, (c) & (d) are the corresponding selected frames with the best quality.

from the same video paragraph of another video. One can note that plots of quality score for slide show frames as given in Fig. 8.12 show much smaller variation compared to that in the case of writing hand frames as given in Fig. 8.11, which is quite expected.

The results of no-reference quality assessment for non-content frames are shown in Fig. 8.13. We notice that frame 8820 offers the best quality of the talking head for the first frame sequence and that frame 6778 does it for the second sequence. The parameters of the quality score evaluation, as given in equation (5.19), used for all test images are $a = 245.9$, $b = 261.9$, $c_1 = 0.0240$, $c_2 = 0.0160$ and $c_3 = 0.0064$, as suggested in [154]. These parameters are the same as used in chapter 5.

The locations of high quality frames, both in the content and non-content segments of the video are used for the selection of video snippets to produce the highlights. The proportion of the different classes of these video snippets are chosen as 1:3 or 1:2:2, as already mentioned and their temporal order is preserved in the synopsis. For media re-creation, about 250 frames, around the high quality frame instants are selected as highlights along with the audio. Thus our algorithm, when applied on a lecture video of 1 hour duration yielded a synopsis of about 2 minutes.

Table 8.3 A comparison of lecture video capsules with their source videos

Video of 1 Hr.	Original size	Size of capsule	Compression factor
Algorithms	2.3 GB	12.6 MB	182.54
Optimization	2.4 GB	13.2 MB	181.82
Random Processes	872 MB	4.6 MB	189.56
Image Processing	2.5 GB	13.4 MB	186.57

Some of the generated capsules are given in the CD-ROM for viewing purposes. On implementation using Matlab 7.8, running on a machine with Intel Quad core processor and 4 GB of RAM, it took around 30 minutes to process a 1 hour video to generate the capsule.

We produced four preview capsules from those videos mentioned in the beginning of this section. Their details are given in Table 8.3. Both the original videos and the resultant capsules are in MPG format. Note that all the capsules are having a playback duration of about 2 minutes. These preview videos will be able to provide only a glimpse of the topics covered by the instructor in the original lecture. These are to be distributed freely in distance education systems. Since there is no existing methods of lecture video preview creation, a comparative study of our method could not be performed. To judge the effectiveness of the prepared capsule, a user has to watch the complete one hour lecture first and then the corresponding capsule. Since this has many practical limitations, is left as a future work.

8.6 Discussions

The development of an efficient algorithm for the synopsis creation of instructional video is discussed in this chapter. As demonstrated, visual quality and content based highlight extraction have been performed differently on separate classes of instructional activities. For non-content classes a no-reference visual quality measure [154] based on blocking effect and blur while for content classes a novel method based on both textual clarity and content quantification have been presented. Measures for clarity were derived from the intensity histogram of the content frame and those for content quantification were obtained from the horizontal projection profile (HPP) of the ink pixels in the content frame. Based on these quality scores, appropriate video snippets are selected and finally these clips are merged in suitable proportions to create a compact video synopsis. It is worth noting that the HPPs of the content frames have been used again and again in this monograph to do various different kind of tasks required for processing of educational videos. This is with the specific purpose that the same HPP data can be used by different modules of the entire video processing system, saving us in computation.

It has been observed that the histogram and lay-out based measures are very powerful in rendering the high quality content frame for clip selection. The philosophy behind adopting this method for defining the *quality* of content frames is that, the amount of writing, its lay-out as well as clarity, all should contribute to the overall quality. Since speech of the instructor does not change much during the lecturing process, and it does not contribute to the highlights of lecture video, it was not used for capsule preparation.

The generated *lecture video capsule* (LVC) can serve as a preview for students in the scenario of the increased use of a variety of educational videos in distance education. A user can have a fast preview of these freely available capsules before taking the decision of attending the course or buying the entire course video. Hence it finds a very potential application in the present world of decentralized education through web-based learning or e-learning.

Chapter 9
Conclusions and Future Directions of Research

The analyses of the methodologies adopted for the repackaging and the subsequent content re-creation of instructional video for technology enhancement in distance education have been provided in previous chapters. The conclusions derived out of this research are summarized in this chapter along with highlighting the novelties, merits and shortcomings in this research area. Some future directions of this research are also included towards the end.

9.1 Conclusions

The preliminary tasks addressed in connection with the repackaging of lecture video are shot detection and recognition. With this objective, two different approaches have been suggested. In the first one, the histogram difference in the intensity space temporally segments the video and an HMM classifier identifies the scenes subsequently. In another approach, the HMM framework designed for the shot classification itself performs the change detection, through a continuous monitoring of the respective likelihood functions using Shiryaev-Robert statistics. This framework is capable of automatically detecting the transitions in scenes, thereby separating them at the transition points so that the individual activities can be efficiently and continuously recognized with a guaranteed minimum delay. Using these methods the instructional activities are classified into three categories: (1) talking head, (2) writing hand and (3) slide show. We give a nomenclature of *content frames* to both the second and the third classes as a whole and *non-content frames* to all frames belonging to the class of talking head.

The first set of investigations in this research is focused on the development of a simple but accurate algorithm for compression efficient, content preserved repackaging of lecture video sequences. In this, computation of visual quality and content based key-frame extraction strategies are applied differently on the separate classes of video segments and a pedagogic representation of these lecture video segments is done effectively. For the selection of key-frames in non-content segments like a talking head, a no-reference perceptual visual quality measure based on the blocking effect and blur and as recommended in [154] is used while for content segments, an HPP based scheme is employed. The Radon transform based skew detection module avoids duplications in content key-frames and yields distinct key-frames. The key-frames for different scenes of the lecture video provide a visual summary and an effective description of the content of the lecture. If the content key-frames suffer from poor resolution, they are super-resolved for improved legibility. A MAP based technique is used to combine frames temporally adjacent to a key-frame and super-resolution of the content frame is achieved. On media synthesis, an estimate of the original lecture video sequence is re-created from these content key-frames, the talking head frames and the associated audio. The reproduced media completely represents the original video in terms of all pedagogic values since it is delivered to the viewer in a semantically organized way, without altering the playback duration of the video. The multimedia summary produced by the content analysis which essentially contains only a few key-frames, a text file with the temporal information as the meta-data and the associated audio can be called an *instructional media package* (IMP). The summarization and the content re-creation of lecture videos as described in Chapter 5 lead to considerable savings in memory when an enormous amount of video data are to be stored. This, in turn, accounts for a fast browsing environment, on a huge volume of lecture video data. Another achievement that can be noted is the reduction in the requirement of transmission bandwidth.

The next part of research is an extension of the above work which aims at developing a method for legibility retentive display of instructional media on miniature display devices that tries to preserve the pedagogic content in the original instructional video. Thousands of video frames have been represented by a few corresponding key-frames which are displayed on the miniature screen as moving key-hole images. The size of the key-hole image is comparable to that of the mobile screen and so only the *region of interest* (ROI) in the key-frame with respect to the audio is displayed with the full resolution. The movement of the key-hole image is controlled by a video meta-data derived from the tracking of the written text in the content segments of the video. This meta-data is derived by using both the HPP and VPP of the ink pixels in the content frames. Originally the meta-data is computed in 50 frame intervals and on display on mobile devices, these are temporally interpolated to yield a faster frame rate for smooth panning. This method does not crop windows from the resident key-frames in the memory. Instead, it selects the rectangular region of interest in accordance with the tracking meta-data and delivers this region on the mobile screen on play back. In the default case, this panning window moves automatically with the tracking meta-data. One may also have a manual control through the touch screen or the selection key on the mobile keypad. Thus the movement of the window can be additionally controlled manually as per wish of the viewer if, for some reason, the hand tracking meta-data performs poorly. Moreover, the visual content delivery is performed not always with the single (fullest) resolution. If the text tracking meta-data variation over a short interval in time is much more than the size of the display window, the algorithm expands the region of interest but then the visual delivery will be at a reduced resolution. This helps in providing a contextual awareness about the written document to the viewer. Hence by using this technique, a legibility retentive visual content delivery is achieved on miniature devices with small displays. This legibility preserved retargeting of instructional video on miniature devices has a great potential of finding applications in distance education systems as now-a-days the end-users are also mobile. It is quite easy to stream the corresponding instructional material directly to the mobile phone for an enhanced outreach. This adds to the goals of distance education by providing the learning materials *anywhere, anytime*. In order to facilitate this goal, an android-based media player called Lec-to-Mobile[1], has been developed and is freely available for downloading from the Google (android) market place.

For the content authentication of the IMP, we use appropriate watermarking techniques in which suitable marks are imperceptibly embedded in the video key-frames and in the audio track. A DCT based spread spectrum scheme is employed for the watermarking of key-frames. The color image watermarking is effected by marking only the blue channel since the human visual system is claimed to be less sensitive to changes in this band. The algorithm embeds the watermark in the midband DCT coefficients of 8×8 blocks of the image. The mid-band is chosen as the embedding region so as to provide an additional resistance to lossy compression techniques, while avoiding any significant modification of the cover image. The robustness of the image watermarking scheme is tested against various attacks like scaling and smoothing. For audio watermarking, we use a multi-segment watermarking scheme based on the audio histogram [160]. This method effectively handles the random cropping and time scale modification attacks, since it uses the features of audio histogram for watermarking which are insensitive to such attacks. The method of decision fusion is employed for achieving high detection accuracies. With the help of watermarking of the IMP, it is expected that content owners will be able to protect their ownership rights, when commercial applications may benefit.

Finally we come up with a novel method of preparing a *lecture video capsule* (LVC) for preview purposes for the users of distance education. For this, measurement of visual quality and content based highlight extraction have been performed differently on separate classes of instructional activities. For non-content classes the same no-reference visual quality metric proposed in [154] is used, while for content classes a novel method based on both textual clarity and content quantification is used. Measures for clarity have been derived from the intensity histogram of the content frame and those for content quantification have been obtained from the horizontal projection profile (HPP) of the ink pixels in the content frame. Based on these quality scores, appropriate short-duration video snippets are selected and finally these clips are merged in suitable proportions to create a compact video synopsis. It has been observed that the histogram and lay-out based measures are very powerful in detecting the high quality content frames for clip selection. The rationale behind adopting this method for defining the *quality* of content frames is that, the amount of writing, its lay-out as well as clarity, all should contribute to the overall quality. Since speech of the instructor need not necessarily be contributing to the

---

[1] Copyright pending.

highlights of lecture video, the method is not based on speech/audio analysis. The generated LVC can serve as a preview for students in the scenario of the increased use of variety of educational videos in distance education. An user can have a fast preview of these freely available capsules before taking the decision of attending the course or buying the entire course video. Hence it finds a very potential application in the present world of decentralized education through web-based learning or e-learning.

The schematic diagram incorporating all of the discussed processing units for instructional video, which contribute to the technology enhancement in distance education systems is shown in Fig. 9.1.

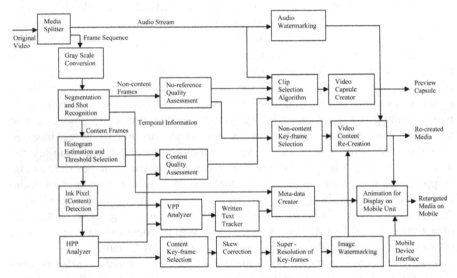

Fig. 9.1 Block schematic for the overall system for the repackaging and the content re-creation for instructional video.

Various aspects of these blocks have been in different chapters of this monograph. They are briefly listed as

a. An optimal shot detection algorithm based on HMM which uses Shiryaev-Robert statistics in Section 4.4,
b. Content analysis of handwritten or electronic slides based on the detected ink pixels in Section 5.2,
c. Key-frame extraction based on the HPP of the ink pixels in the content segments in instructional video in Section 5.2,
d. Skew correction of content key-frames based on Radon transform in Section 5.3,
e. Effective content re-creation of instructional video from *instructional media package* (IMP) in Section 5.6,

f. Tracking of written text in the instructional video based on both the HPP and VPP of ink pixels in Section 6.2,
g. Legibility retentive adaptation of instructional media on miniature displays in Section 6.3,
h. Quality assessment of content frames in instructional video based on intensity histogram as well as lay-out features in Section 8.2,
i. Production of a *lecture video capsule* (LVC) for preview purposes in Section 8.4.

Some of the limitations of this approach due to the occasional violation of the assumptions followed are now discussed. This work assumes only three classes of instructional activities, so the algorithms fail for instructional videos with other classes (such as shots depicting students asking questions, etc.) of instructional activities. Also those medium distance shots in which an instructor appears along with writing on a black/white board is not addressed here. This work focuses on instructional videos specially prepared for distance education, which usually have a sequence of closely projected handwritten or electronic slides. Also the picture-in-picture (PIP) technique sometimes incorporated in lecture videos, by which a small talking head field is superposed on the non-content areas of an electronic slide or handwritten frame is not addressed in this work. The other categories like discussion sessions, imported videos, etc., have not been considered in this study. The text tracking can be effectively performed with the use of HPP and VPP in the case of handwritten slides only. For electronic slide shows, as there is often no sequential writing or display of characters within a slide, the only possibility of finding the *region of interest* (ROI) in the frame is by processing the associated audio, which has not been addressed here. In such a case, it requires a combination of a speech to text converter and an optical character recognizer (OCR) to find the ROI in the instructional video for an intelligent display on miniature mobile devices.

9.2 Future directions of research

Our effort in this monograph has been directed towards developing a functionally complete, working system for the content re-creation of instructional video. It did use some of the existing techniques which have been studied well in literature. However, its performance evaluation and user study could be done in the next phase of research. With that in mind, a media player, called Lec-to-Mobile, has been released to the public and it is hoped that appropriate feedback from the users of this player would help us in further improving the system.

Considering the limitations of assumptions used in this thesis, further developments could be performed as an extension to the various methodologies proposed. One such extension is increasing the number of classes of instructional activities. A lecture video may occasionally contain discussion sessions as well as imported video segments. One can incorporate these additional classes for improving the generality of the scheme. Another area of future work is tackling the PIP technique occasionally found in instructional videos. Shot detection or key-frame extraction may occasionally fail for the mixed category of PIP frame sequences since the visual content belongs to both textual and non-textual (talking head) regions. Hence techniques for handling frames with PIP have to be appropriately developed. Another extension of work is the legibility retentive retargeting of IMP of slide show videos on mobile devices. As already mentioned, for electronic slide shows, the ROI in the slide could be identified by processing the speech signal. That is, a combination of OCR and a speech recognizer can link the speech to different areas in the electronic slide. Hence the slide show content could be legibility retentively retargeted on miniature devices. Another interesting topic for future research would be related to streaming the instructional media under different application scenarios.

Appendix A
Details of the Instructional Media Package (IMP)

A.1 Components of the IMP

Image files: A few key frames from the video, generally in JPEG format.

Audio file: The audio file of the instructor in appropriate audio file format (WAV, MP3 etc.).

A text file: The file containing video meta-data, in XML format.

A.2 Content of the XML file

The XML file basically should contain the following information :

(i) Name and type (content or non-content) of the image file to be currently rendered on mobile screen.
(ii) The time for which a particular frame is to be displayed.
(iii) x-y co-ordinates of the written text in the case of content frame.
(iv) Name of the associated audio file.

The corresponding XML file looks like :

```
<LECTURE>
    <FRAME type = "NC">
        <NAME>keyframe1.jpg</NAME>
        <TIME>5000</TIME>
    </FRAME>
    <FRAME type= "C">
        <NAME>keyframe4.jpg</NAME>
        <TIME>2000</TIME>
        <X>39</X>  <Y>11</Y>
        <X>194</X>  <Y>135</Y>
        <X>254</X>  <Y>154</Y>
```

```
        <X>...</X> <Y>...</Y>
      </FRAME>
      <SOUND>sound1.wav</SOUND>
    </LECTURE>
```

A.3 Tags used by the XML file

Tags used in XML File are :

LECTURE : The root node.

FRAME : The information about type of key frame to be displayed on mobile screen.

NAME : The name of the image to be displayed.

TIME : The time duration for which a key frame or key-hole image is to be displayed.

X : The x co-ordinate of the center of the key-hole image.

Y : The y co-ordinate of the center of the key-hole image.

SOUND : The name of the sound file to be played in sequence.

Appendix B
Design Aspects of Lec-to-Mobile

Lec-to-Mobile is a mobile multimedia player designed to deliver the contents of the IMP to the user. It renders the *feel* of a video to the mobile user, when the IMP is played by it. In this respect, it is a *reader* used to access the contents of IMP on a mobile environment. Since the android based player is already available for free downloading as mentioned in Section 6.4, we provide an alternate implementation using Windows mobile OS.

B.1 Implementation details

Programming Platform :   Microsoft Visual Studio 2008.
Programming Language: Visual C++.
Test Hardware: Motorola Q GSM.
Software Development Kit (SDK):   Windows mobile Smartphone 5.0 SDK.
Additional Support :   MSDN library for visual studio 2008-ENU.
Software used: Windows Mobile Device Center 6.1.

The Program is designed as a Win32 Application to run on a Smartphone as well as on Pocket PC using MSDN (Microsoft Developer Network) library. In order to develop our application in Visual studio 2008 we have used MSDN library for visual studio 2008-ENU. MSDN Library is a library of official technical documentation content intended for developers developing in Microsoft Windows. The Windows Mobile 5.0 SDK extends Visual Studio 2008 so as to write managed and native application software targeting Windows Mobile 5.0 based Smartphone and Pocket PC devices. This includes Windows Mobile 5.0 based Smartphone Device Emulator images and skin files, Windows Mobile 5.0 based Pocket PC Device Emulator images and skin files Headers, Libraries, IDL files, and managed reference assemblies, and Sample Code.

In order to transfer information in and out of the emulator or real device as well to install and remove programs we use ActiveSync for Windows XP laptops and

PCs or Windows Mobile Device Center for Windows Vista (all versions) and Windows 7. These synchronize e-mail, Calendar, Contacts, Tasks of the device with the system. The emulator is synchronized after cradling it with help of Device Emulator Manager already installed in Visual Studio 2008. The Test Hardware used is Motorola Q GSM which uses Windows Mobile Smartphone 5 SDK.

B.2 Development of the media player

B.2.1 Reading data from xml file

The first step to re-create the media on mobile is to read and interpret the data stored in xml file. We have used C^{++} XML Parser CMarkup to extract information from the xml file provided from the server side. We make use of the following CMarkup methods while achieving our target:

Load: Populates the CMarkup object from a file and parses it. It takes path of the meta-data xml file as an input. It returns true if the file is successfully loaded.

ResetPos: Clears the current position so that there is no current main position or child position and the current parent position is reset to the top of the document.

FindElem: Goes forward from the current main position to the next matching sibling element. If there is no next sibling element it returns false and leaves the main position where it was. It has an optional argument which allows you to specify a tag name or path.

FindChildElem: Locates the first child element under the main position and makes it the child position. If there is a child element, it moves to the next child element. If there is no child element or no next child element, it returns false and leaves the child position where it was.

IntoElem: Makes the main position element the parent element. If there is no main position it returns false and does not affect the current position at all.

AddElem: Is called to add an element after the main position. The tag name of the element to be added is passed as an argument. The data value is optional. The method returns true if the element is added successfully and the main position points to it.

GetAttrib: To get the string value of the attribute in the main position element. If there is no current position, it returns an empty string.

GetData: Returns the string data value of the main position element or node. It returns an empty string if there is no data value, or there is no current position, or the main position element contains child elements.

GetChildData: Returns the string data value of the child position element. It returns an empty string if there is no data, or there is no child position or the child position element contains descendants.

Save: Write the document to file. It takes path of the xml file to be created as an
    input. It will overwrite the file if it exists and create it if it does not. It returns true
    if the file was successfully written.

B.2.2 Displaying key-frames on mobile screen

Once we have read and interpreted the data stored in xml file, the next step is to
display an image on mobile screen. The algorithm allows displaying the part of the
image as well as the entire image on the mobile screen. For windows mobile we
have used Imaging API which is the part of Graphical Device Interface (GDI). This
API provides information about adding support for compressed still images to OS
design and its application. We have used the following two interfaces included in
Imaging API:

IImagefactory

This interface is used to create bitmaps and images and to manage image encoders
and decoders. In order to display image we use its following function:
`IImagingFactory::CreateImageFromFile`: This method lets an application create a decoded image object from a file. It takes as the argument the path of
the source image and a pointer to an IImage interface pointer.

IImage

This is the basic interface to an image object. It allows applications to do the following:

(i) Display the image onto a destination graphics context.
(ii) Push image data into an image sink.
(iii) Access image properties and meta-data.

    In order to draw image on mobile screen we used its following function:
`IImage::Draw`: This method displays the image onto the specified area of a destination graphics context. It takes the following as the arguments

(i) `hdc`: HDC (handle to a device context) value of the graphics context that
    receives output from this method.
(ii) `dstRect`: A pointer to a RECTANGLE defining the portion of the display
    area within the graphics context that receives the output from this method.
(iii) `srcRect`: An optional pointer to a RECTANGLE that specifies, in 0.01 mm
    units, the portion of the image to be drawn in dstRect.

B.2.2.1 Displaying non-content key-frames

The frames which consist of talking head, audience (discussion) or the face of the instructor are referred to as non-content key-frames. The representative key-frame of the talking head sequence is just scaled down and displayed on mobile screen as they account only for some sort of situational awareness to the viewer. They are displayed for the time duration for which the corresponding segment appears in the original video. The xml data corresponding to non content key frames consist of the name of the image to be displayed, its type specified as "NC" in attribute and the time for which it has to be displayed.

In order to display non-content key frame we set the srcRect of iimage::draw function as NULL and it takes the entire image, scales it and shows it on mobile screen.

B.2.2.2 Displaying content key-frames

The writing hand and slide show frames, called content frames, are to be displayed by virtual panning of the original key-frame with a window size equal to the size of the miniature screen in accordance with a meta-data. Moreover, if the content frame has a prolonged idle hand or it is a clean frame for some time, we display the whole key-frame, resized to fit to the mobile screen. The xml data corresponding to content key-frames consist of the name of the image to be displayed, its type specified as "C" in attribute, x and y co-ordinates and time interval between two meta-data points.

In order to display content key-frame we set the srcRect of iimage::draw function equal to the rectangle which specify the key-hole image and display it on mobile screen.

B.2.3 Playing audio

The audio file is played in background in synchronism with the image display. Our system is capable of buffering a sound file of size less than 10 MB at a time. For file of size greater than 10 MB we can break them into files of size less than 10 MB and play them sequentially. We have used wave API to play our sound files one after the other. We have used the following function of wave API:

waveOutOpen: Opens a specified waveform output device for playback.
waveOutPrepareHeader: Prepares a waveform data block for playback. The application can use the buffer repeatedly without additional processing by the driver or the OS.
waveOutWrite: Sends a data block to the specified waveform output device.
waveOutUnprepareHeader: Cleans up the preparation performed

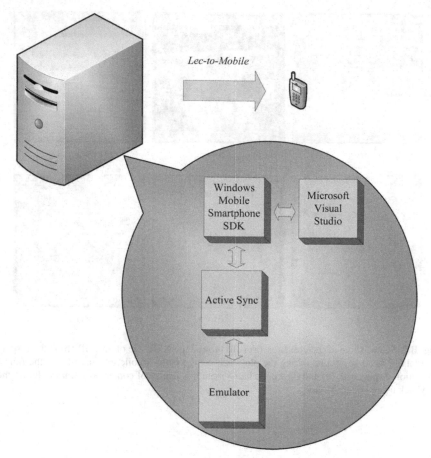

Fig. B.1 Block schematic for the development of Lec-to-Mobile.

by `waveOutPrepareHeader`. The function must be called after the device driver is finished with a data block.
`waveOutClose`: Closes the specified waveform output device.

The overall process of creating Lec-to-Mobile is given in Fig. B.1

B.3 Emulator results

Lec-to-Mobile is developed and tested using the emulator, after which is deployed in the mobile device as a player for the IMP which is already stored in the mobile memory. Some screen shots of the emulator are shown in Fig. B.2.

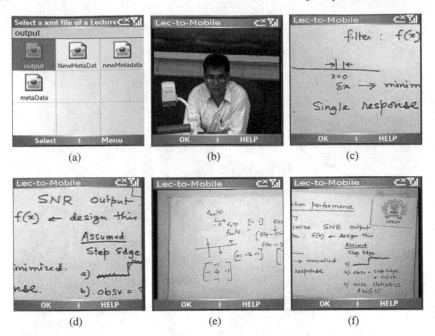

Fig. B.2 Screen shots of the emulator. (a) is to browse the XML file of an IMP, (b) is the display of a talking head key-frame during the playback using Lec-to-Mobile, (c) and (d) are the fullest resolution display and, (e) and (f) are the half-resolution display of content key-frames during the playback using Lec-to-Mobile.

References

1. I. Ahmad, X. Wei, Y. Sun, and Y.-Q. Zhang. Video transcoding: an overview of various techniques and research issues. *IEEE Transactions on Multimedia*, 7(5):793–804, October 2005.
2. N. Ahmidi and R. Safabakhsh. A novel dct-based approach for secure color image watermarking. *Proceedings of the International Conference on Information Technology: Coding and Computing*, 2:709–713, April 2004.
3. A. Al-Gindy, H. Al-Ahmad, R. Qahwaji, and A. Tawfik. Watermarking of colour images in the dct domain using y channel. *Proceedings of the IEEE/ACS International Conference on Computer Systems and Applications*, pages 1025–1028, May 2009.
4. A. M. Alattar. Reversible watermark using the difference expansion of a generalized integer transform. *IEEE IEEE Transactions on Image Processing*, 13(8):1147–1156, August 2004.
5. P. Anandan, M. Irani, R. Kumar, and J. Bergen. Video as an image data source: efficient representations and applications. *Proceedings of the International conference on Image Processing*, 1:318–321, October 1995.
6. Avanindra and S. Chaudhuri. Robust detection of skew in document images. *IEEE Transactions on Image Processing*, 6(2):344–349, February 1997.
7. M. Barni, F. Bartolini, V. Cappellini, and A. Piva. A dct-domain system for robust image watermarking. *Signal Processing*, 66(3):357–372, May 1998.
8. M. Barni, F. Bartolini, and A. Piva. Multichannel watermarking of color images. *IEEE Transactions on Circuits and Systems for Video Technology*, 12(3):142–156, March 2002.
9. P. Bassia, I. Pitas, and N. Nikolaidis. Robust audio watermarking in the time domain. *IEEE Transactions on Multimedia*, 3(2):232–241, June 2001.
10. P. Bassia, I. Pitas, and N. Nikolaidis. Robust audio watermarking in the time domain. *IEEE Transactions on Multimedia*, 3(2):232–241, June 2001.
11. W. Bender, D. Gruhl, N. Morimoto, and A. Lu. Techniques for data hiding. *IBM systems journal*, 35(3/4):313–336, 1996.
12. N. Bjork and C. Christopoulos. Video transcoding for universal multimedia access. *Proceedings of the ACM International Conference on Multimedia*, 1:75–79, 2000.
13. M. Bonuccellit, F. Lonetti, and F. Martelli. Temporal transcoding for mobile video communication. *Proceedings of the Second Annual International Conference on Mobile and Ubiquitous Systems: Networking and Services*, pages 502–506, July 2005.
14. J. Boreczky and L. Wilcox. A hidden markov model framework for video segmentation using audio and image features. *Proceedings of IEEE International Conference on Acoustics, Speech and Signal Processing*, 6:3741–3744, 1998.
15. M. Brand and V. Kettnaker. Discovery and segmentation of activities in video. *IEEE Transactions on Pattern Analysis and Machine Intelligence*, 22(8):844–851, August 2000.
16. J. Brandt and L. Wolf. Adaptive video streaming for mobile clients. *Proceedings of the 18th International Workshop on Network and Operating System Support for Digital Audio and Video*, pages 113–114, 2008.
17. J. Brandt, L. Wolf, and P. Halvorsen. Multidimensional transcoding for adaptive video streaming. *ACM SIGMultimedia*, 1(4):16–17, December 2009.
18. Carson, Belongie, H. Greenspan, and J. Malik. Blobworld: Image segmentation using expectation-maximization and its application to image querying. *IEEE Transactions on Pattern Analysis and Machine Intelligence*, 24(8):1026–1038, August 2002.
19. Z. Cernekova, I. Pitas, and C. Nikou. Information theory-based shot cut/fade detection and video summarization. *IEEE Transactions on Circuits and Systems for Video Technology*, 16(1):82–91, January 2006.
20. H. S. Chang, S. Sull, and S. U. Lee. Efficient video indexing scheme for content-based retrieval. *IEEE Transactions on Circuits and Systems for Video Technology*, pages 1269–1279, December 1999.
21. P. Chang, M. Han, and Y. Gong. Extract highlights from baseball game video with hmm. *Proceedings of the IEEE International Conference on Image Processing*, 2002.

22. V. T. Chasanis, A. C. Likas, and N. P. Galatsanos. Scene detection in videos using shot clustering and sequence alignment. *IEEE Transactions on Multimedia*, 11(1):89–100, January 2009.

23. S. Chaudhuri and M. V. Joshi. *Motion-free super-resolution*. Springer Publishers, New York, USA, 2005.

24. H.-Y. Chen, G.-Y. Chen, and J.-S. Hong. Design of a web-based synchronized multimedia lecture system for distance education. *Proceedings of the IEEE International Conference on Multimedia Computing and Systems*, 2:887–891, July 1999.

25. M. Chen, E. K. Wong, N. Memon, and S. Adams. Recent developments in document image watermarking and data hiding. *Proceedings of the SPIE Conference on Multimedia Systems and Applications*, 4:1025–1028, 2001.

26. C. Choudary and T. Liu. Summarization of visual content in instructional videos. *IEEE Transactions on Multimedia*, 9(7):1443–1455, November 2007.

27. P. Clark and M. Mirmehdi. Finding text regions using localised measures. *Proceedings of the 11th British Machine Vision Conference*, 2000.

28. D. C. Coll and G. H. Choma. Image activity characteristics in broadcast television. *IEEE Transactions on Communications*, pages 1201–1206, Ocober 1976.

29. I. J. Cox, J. Kilian, F. T. Leighton, and T. Shamoon. Secure spread spectrum watermarking for multimedia. *IEEE Transactions on Image Processing*, 6(12):1673–1687, December 1997.

30. R. Cucchiara, C. Grana, and A. Prati. Semantic video transcoding using classes of relevance. *International Journal of Image and Graphics*, 2003.

31. S. Dagtas and M. Abdel-Mottaleb. Extraction of tv highlights using multimedia features. *Proc. of fourth IEEE workshop on Multimedia signal processing*, 2001.

32. A. Dempster, N. Laird, and D. Rubin. Maximum likelihood from incomplete data via the em algorithm. *Journal of Royal Statistical Society Ser. B*, 39(1):1–38, 1977.

33. J.-X. Dong, P. Dominique, A. Krzyyzak, and C. Y. Suen. Cursive word skew or slant corrections based on radon transform. *Proceedings of the Eighth International Conference on Document Analysis and Recognition*, pages 478–483, 2005.

34. C. Dorai, V. Oria, and V. Neelavalli. Structuralizing educational videos based on presentation content. *Proceedings of the International Conference on Image Processing (ICIP)*, 2:1029–1032, September 2003.

35. B. Dugonik, Z. Brezocnik, and M. Debevc. Video production for distance education. *Proceedings of the 24th International Conference on Information Technology Interfaces (ITI)*, 1:141–145, 2002.

36. A. Ekin, A. M. Tekalp, and R. Mehrotra. Automatic soccer video analysis and summarization. *IEEE Transactions on Image Processing*, 12(1):796–807, July 2003.

37. X. Fan, X. Xie, H.-Q. Zhou, and W.-Y. Ma. Looking into video frames on small displays. *Proceedings of the eleventh ACM international conference on Multimedia*, pages 247–250, July 2003.

38. S. Farsiu, M. D. Robinson, M. Elad, and P. Milanfar. Fast and robust multiframe super resolution. *IEEE Transactions on Image Processing*, 13(10):1327–1344, October 2004.

39. G. A. Fink, M. Wienecke, and G. Sagerer. Video based on-line handwriting recognition. *Proceedings of the sixth International Conference on document analysis and recognition*, pages 226–230, September 2001.

40. R. B. Gandhi, J. C. Metcalf, J. K. O'Connel, and M. C. Taskiran. Method for intelligently creating, consuming and sharing video content on mobile devices. *Patent document, International Pub. No. WO 2009/042340 A2*, 2009.

41. U. Gargi, R. Kasturi, and S. H. Strayer. Performance characterization of video-shot-change detection methods. *IEEE Transactions on Circuits and Systems for Video Technology*, 10(1):1–13, February 2000.

42. Y. Gong and X. Liu. Generating optimal video summaries. *Proceedings of the IEEE International conference on Multimedia and Expo*, pages 1559–1562, February 2000.

43. R. C. Gonzalez and R. E. Woods. *Digital Image Processing*. Addison-Wesley Publishing Company, USA, 1993.

44. L. Gupta and T. Sortrakul. A gaussian-mixture-based image segmentation algorithm. *Elsevier Pattern Recognition*, 31(3):315—325, 1998.

45. Han and Kweon. Shot detection combining bayesian and structural information. *Proceedings of SPIE, the International Society for Optical Engineering*, 4315:509–516, 2001.

46. Hanjalic and Zhang. Optimal shot boundary detection based on robust statistical models. *Proceedings of the IEEE International Conference on Multimedia Computing and Systems*, 2:710–714, 1999.

47. A. Hanjalic. Shot-boundary detection: Unraveled and resolved ? *IEEE Transactions on Circuits and Systems for Video Technology*, 12(2), February 2002.

48. M. M. Hannuksela, Y.-K. Wang, and M. Gabbouj. Isolated regions in video coding. *IEEE Transactions on Multimedia*, 6(2):259–267, April 2004.

49. D. A. Harris and C. Krousgrill. Distance education: New technologies and new directions. *Proceedings of the IEEE*, 96(6):917–930, June 2008.

50. F. Hartung and M. Kutter. Multimedia watermarking techniques. *Proceedings of the IEEE*, 87(7):1079–1107, July 1999.

51. W. J. Heng and Q. Tan. Content enhancement for e-learning lecture video using foreground/background separation. *Proceedings of the IEEE Workshop on Multimedia Signal Processing*, pages 436–439, December 2002.

52. J. R. Hernandez, M. Amado, and F. Perez-Gonzalez. Dct-domain watermarking techniques for still images: detector performance analysis and a new structure. *IEEE Transactions on Image Processing*, 9(1):55–68, January 2000.

53. K. Hoganson and D. Lebron. Lectures for mobile devices, evaluating "ipod/pda-casting" technology and pedagogy. *Proceedings of the Fourth International Conference on Information Technology*, pages 323–328, April 2007.

54. G. Hua, C. Zhang, Z. Zhang, Z. Liu, and Y. Shan. Video retargeting. *Patent document, Pub. No. US 2009/0251594 A1*, 2009.

55. C.-L. Huang and B.-Y. Liao. A robust scene-change detection method for video segmentation. *IEEE Transactions on Circuits and Systems for Video Technology*, 11(12):1281–1288, December 2001.

56. J.-N. Hwang, T.-D. Wu, and C.-W. Lin. Dynamic frame-skipping in video transcoding. *IEEE Second Workshop on Multimedia Signal Processing*, pages 616–621, December 1998.

57. M. Irani, P. Anandan, and S. Hsu. Mosaic based representations of video sequences and their applications. *Proceedings of the Fifth International Conference on Computer Vision*, pages 605–611, June 1995.

58. A. K. Jain, R. P. W. Dulin, J. Mao, and S. Member. Statistical pattern recognition: A review. *IEEE Transactions on Pattern Analysis and Machine Intelligence*, 22:4–37, 2000.

59. F. Jiang and J. Xiao. Content-aware image and video resizing by anchor point sampling and mapping. *Patent document, Pub. No. US 2010/0124371 A1*, 2010.

60. N. Johnson and S. Katezenbeisser. A survey of steganographic techniques. *Information Techniques for Steganography and Digital Watermarking*, pages 43–75, December 1999.

61. R. Kapoor, D. Bagai, and T. S. Kamal. A new algorithm for skew detection and correction. *Elsevier Pattern Recognition Letters*, pages 1215–1229, 2004.

62. J.-H. Kim, J.-S. Kim, and C.-S. Kim. Image and video retargeting using adaptive scaling function. *Proceedings of the 17th European signal processing conference*, pages 819–823, 2009.

63. J.-S. Kim, J.-H. Kim, and C.-S. Kim. Adaptive image and video retargeting technique based on fourier analysis. *Proceedings of the IEEE Conference on Computer Vision and Pattern Recognition*, pages 1730–1737, June 2009.

64. J. Kittler and J. Illingworth. On threshold selection using clustering criteria. *IEEE Transactions on SMC*, 15:652–655, 1985.

65. J. Kittler and J. Illingworth. Minimum error thresholding. *Pattern Recognition, Pergamon Press Ltd, U.K.*, 19(1):41–47, 1986.

66. S. K. Krishna, R. Subbarao, S. Chaudhuri, and A. Kumar. Parsing news video using integrated audio-video features. In *First International Conference on Pattern Recognition and Machine Intelligence (PReMI)*, pages 538–543, 2005.

67. L. I. Kuncheva. A theoretical study on six classifier fusion strategies. *IEEE Transactions on Pattern Analysis and Machine Intelligence*, 24(2):281–286, February 2002.

68. M. Kutter, F. Jordan, and F. Bossen. Digital signature of colour images using amplitude modulation. *Storage and retrieval of image and video databases V, SPIE*, 3022:518–526, February 1997.

69. G. Langelaar, I. Setyawan, and R. L. Lagendijk. Watermarking digital image and video data : A state-of-the-art overview. *IEEE Signal Processing Magazine*, 17(5):20–46, September 2000.

70. S.-W. Lee and D.-S. Ryu. Parameter-free document layout analysis. *IEEE Transactions on Pattern Analysis and Machine Intelligence*, 23(11):1240–1256, November 2001.

71. W. Li, X. Xue, and P. Lu. Localized audio watermarking technique robust against time-scale modification. *IEEE Transactions on Multimedia*, 8(1):60–69, February 2006.

72. Y. Li and C. Dorai. Instructional video content analysis using audio information. *IEEE Transactions on Audio, Speech and Language Processing*, 14(6):2264–2274, November 2006.

73. Y. Li, Y. Tian, J. Yang, L.-Y. Duan, and W. Gao. Video retargeting with multi-scale trajectory optimization. *Proceedings of the international conference of multimedia information retrieval*, pages 45–54, 2010.

74. Y. Li, T. Zhang, and D. Tretter. An overview of video abstraction techniques. *Technical Report, Hewlett and Packard Labs, Hewlett-Packard Company*, July 2001.

75. Z. Li, P. Ishwar, and J. Konrad. Video condensation by ribbon carving. *IEEE Transactions on Image Processing*, 18(11):2572–2583, November 2009.

76. Y. Liang and Y.-P. Tan. A new content-based hybrid video transcoding method. *Proceedings of the International Conference on Image Processing*, 1:429–432, 2001.

77. C.-W. Lin and Y.-R. Lee. Fast algorithms for dct-domain video transcoding. *Proceedings of the International Conference on Image Processing*, 1:421–424, 2001.

78. M. Lin, M. Chau, J. F. N. Jr., and H. Chen. Segmentation of lecture videos based on text: A method combining multiple linguistic features. *Proceedings of the 37th Hawaii International Conference on System Sciences*, 2004.

79. F. Liu and M. Gleicher. Video retargeting : Automatic pan and scan. *Proceedings of the 14th annual ACM international conference on Multimedia*, pages 241–250, 2006.

80. H. Liu, X. Kong, X. Kong, and Y. Liu. Content based color image adaptive watermarking scheme. *IEEE International Symposium on Circuits and Systems*, 2:41–44, May 2001.

81. L. Liu, Y. Dong, X. Song, and G. Fan. An entropy-based segmentation algorithm for computer-generated document images. *Proceedings of IEEE International Conference on Image Processing*, 1:541–544, September 2003.

82. T. Liu and C. Choudary. Content-aware streaming of lecture videos over wireless networks. *Proceedings of the IEEE Sixth International Symposium on Multimedia Software Engineering*, pages 458–465, December 2004.

83. T. Liu and C. Choudary. Content extraction and summarization of instructional videos. *Proceedings of the International Conference on Image Processing*, pages 149–152, October 2006.

84. T. Liu and J. R. Kender. Rule-based semantic summarization of instructional videos. *Proceedings of IEEE International Conference on Image Processing*, 1:601–604, 2002.

85. T. Liu and J. R. Kender. Lecture videos for e-learning: Current researches and challenges. *Proceedings of the IEEE Sixth International Symposium on Multimedia Software Engineering*, pages 574–578, December 2004.

86. T. Y. Lui and E. Izquierdo. Scalable object-based image retrieval. *Proceedings of IEEE International Conference on Image Processing*, 3:501–504, September 2003.

87. H. S. Malvar and D. A. F. Florencio. Improved spread spectrum: a new modulation technique for robust watermarking. *IEEE Transactions on Signal Processing*, 51(4):898–905, April 2003.

88. C. D. Manning, P. Raghavan, and H. Schutze. *An Introduction to Information Retrieval*. Cambridge University Press, 2009.

89. P. Marziliano, F. Dufaux, S. Winkler, and T. Ebrahimi. A no-reference perceptual blur metric. *Proceedings of IEEE International Conference on Image Processing*, 3:57–60, 2002.

90. T. K. Moon. The expectation-maximization algorithm. *IEEE Signal Processing Magazine*, pages 47–60, November 1996.
91. M. E. Munich and P. Perona. Visual input for pen-based computers. *Proceedings of the International Conference on Pattern Recognition*, 3:33–37, 1996.
92. G. N. Murthy and S. Iyer. Study element based adaptation of lecture videos to mobile devices. *Proceedings of the National Conference on Communications*, pages 1–5, January 2010.
93. H.-M. Nam, K.-Y. Byun, J.-Y. Jeong, K.-S. Choi, and S.-J. Ko. Low complexity content-aware video retargeting for mobile devices. *IEEE Transactions on Consumer Eletronics*, 56(1):182–189, February 2010.
94. C.-W. Ngo, F. Wang, and T.-C. Pong. Structuring lecture videos for distance learning applications. *Proceedings of the IEEE Fifth International Symposium on Multimedia Software Engineering*, pages 215–222, December 2003.
95. N. Nikolaidis and I. Pitas. Robust image watermarking in the spatial domain. *Signal Processing*, 66(3):385–403, May 1998.
96. F. Niu and M. Abdel-Mottaleb. Hmm-based segmentation and recognition of human activities from video sequences. *IEEE International conference on Multimedia and Expo*, July 2005.
97. T. Nuriel and D. Malah. Region-of-interest based adaptation of video to mobile devices. 4th *International Symposium on Communications, Control and Signal Processing*, pages 1–6, March 2010.
98. M. Osadchy and D. Keren. A rejection-based method for event detection in video. *IEEE Transactions on Circuits and Systems for Video Technology*, 14(4):534–541, April 2004.
99. N. Otsu. A threshold selection method from grey-level histograms. *IEEE Transactions on Systems, Man and Cybernetics*, 9(1):62–66, 1979.
100. N. R. Pal and S. K. Pal. Object-background segmentation using new definition of entropy. *IEEE Proceedings*, 136(4):284–295, July 1989.
101. N. R. Pal and S. K. Pal. Entropy: A new definition and its applications. *IEEE Transactions on Systems, Man and Cybernetics*, 21(5):1260–1270, October 1991.
102. A. Papoulis and S. U. Pillai. *Probability, random variables and stochastic processes*. Tata McGraw-Hill, New Delhi, India, fourth edition, 2002.
103. S. C. Park, M. K. Park, and M. G. Kang. Super-resolution image reconstruction : a technical overview. *IEEE Signal Processing Magazine*, pages 21–36, 2003.
104. N. V. Patel and I. K. Sethi. Video shot detection and characterization for video databases. *Pattern Recognition*, 30(4):583–592, April 1997.
105. J. Peng and Q. Xiaolin. Keyframe-based video summary using visual attention clues. *IEEE Multimedia*, 17(2):64–73, June 2010.
106. F. A. P. Petitcolas. Watermarking schemes evaluation. *IEEE Signal Processing Magazine*, 17(5):58–64, September 2000.
107. R. W. Picard. Light-years from lena: video and image libraries of the future. *Proceedings of the IEEE International Conference on Image Processing*, 1:310–313, October 1995.
108. A. Piva, M. Barni, F. Bartolini, and V. Cappellini. Dct-based watermark recovering without resorting to the uncorrupted original image. *Proceedings of the International Conference on Image Processing*, 1:520–523, October 1997.
109. C. I. Podilchuk and E. J. Delp. Digital watermarking: algorithms and applications. *IEEE Signal Processing Magazine*, 18(4):33–46, July 2001.
110. M. Pollak. Optimal detection of a change in distribution. *Annals of Statistics*, 13:206–227, 1985.
111. Y. Pritch, S. Ratovitch, A. Hendel, and S. Peleg. Clustered synopsis of surveillance video. *Proceedings of the Sixth IEEE International Conference on Advanced Video and Signal Based Surveillance*, pages 195–200, September 2009.
112. L. R. Rabiner. A tutorial on hidden markov models and selected applications in speech recognition. *Proceedings of IEEE*, 77:257–286, 1989.
113. L. R. Rabiner and B.-H. Juang. An introduction to hidden markov models. *IEEE ASSP Magazine*, January 1986.

114. A. Rav-Acha, Y. Pritch, and S. Peleg. Making a long video short : Dynamic video synopsis. *Proceedings of the IEEE Computer Society Conference on Computer Vision and Pattern Recognition (CVPR)*, 1:435–441, June 2006.

115. R. A. Redner and H. F. Walker. Mixture densities, maximum likelihood and the em algorithm. *SIAM Review*, 26(2):195–239, April 1984.

116. S. Repp and M. Meinel. Semantic indexing for recorded educational lecture videos. *Proceedings of the Fourth Annual IEEE International Conference on Pervasive Computing and Communications Workshops*, pages 240–245, March 2006.

117. N. Robertson and I. Reid. A general method for human activity recognition in video. *Journal of Computer Vision and Image Understanding*, 104(2):232–248, November 2006.

118. M. Rubinstein, A. Shamir, and S. Avidan. Improved seam carving for video retargeting. *International Conference on Computer Graphics and Interactive Techniques(ACM SIGGRAPH)*, (16), 2008.

119. Y. Rui, A. Gupta, and A. Acero. Automatically extracting highlights for tv baseball programs. *Eighth ACM International Conference on Multimedia*, pages 105–115, 2000.

120. Y. Rui, T. S. Huang, and S. Mehrotra. Content-based image retrieval with relevance feedback in mars. *Proceedings of the International Conference on Image Processing*, 2:815–818, October 1997.

121. Y. Rui, T. S. Huang, M. Ortega, and S. Mehrotra. Relevance feedback: A power tool for interactive content based image retrieval. *IEEE Transactions on Circuits and Systems for Video Technology*, 8(5):644–655, September 1998.

122. E. Sahouria and A. Zakhor. Content analysis of video using principal components. *Proceedings of the International Conference on Image Processing*, 3:541–545, October 1998.

123. B. Salt. Statistical style analysis of motion pictures. *Proceedings of Film Quarterly*, 28:13–22, 1973.

124. G. Salton. *The SMART retrieval system - Experiments in automatic document processing*. Prentice Hall, 1971.

125. G. Salton and M. J. McGill. *Introduction to modern information retrieval*. McGraw Hill Book Company, 1983.

126. E. Sayrol, J. Vidal, S. Cabanillas, and S. Santamaria. Optimum watermark detection in color images. *Proceedings of the International Conference on Image Processing*, 2:231–235, October 1999.

127. C. Seibel and F. Raynal. Stirmark benchmar: Audio watermarking at tracks. *Proceedings of the international conference on information technology: coding and computing*, page 49, 2001.

128. K. Seo, J. Ko, I. Ahn, and C. Kim. An intelligent display scheme of soccer video on mobile devices. *IEEE Transactions on Circuits and Systems for Video Technology*, 17(10):1395–1401, October 2007.

129. A. Shamir and S. Avidan. Seam carving for media retargeting. *Communications of the ACM*, 52(1):77–85, January 2009.

130. A. Shamir and O. Sorkine. Visual media retargeting. *Proceedings of the international conference on computer graphics and interactive technologies*, (11), September 2009.

131. H. R. Sheikh, A. C. Bovik, and L. K. Cormack. No-reference quality assessment using natural scene statistics: Jpeg 2000. *IEEE Transactions on Image Processing*, 14(12):1918–1927, December 2005.

132. A. N. Shiryaev. On optimum methods in quickest detection problems. *Theory of probability and applications*, 8:22–46, 1963.

133. A. N. Shiryaev. Optimal stopping rules. *Springer-Verlag, NY*, 1978.

134. L. S. Silva and J. Scharcanski. Video segmentation based on motion coherence of particles in a video sequence. *IEEE Transactions on Image Processing*, 19(4):1036–1049, April 2010.

135. Simakov, Caspi, Shechtman, and M. Irani. Summarizing visual data using bidirectional similarity. *Proceedings of the IEEE Conference on Computer Vision and Pattern Recognition*, pages 1–8, June 2008.

136. X. Song and G. Fan. Joint key-frame extraction and object segmentation for content-based video analysis. *IEEE Transactions on Circuits and Systems for Video Technology*, 16(7):904–914, July 2006.

137. S. N. Srihari and V. Govindaraju. Analysis of textual images using the hough transform. *Machine Vision and Applications*, 2(3):142–153, June 1989.

138. J. K. Su, F. Hartung, and B. Girod. Digital watermarking of text, image, and video documents. *Elsevier Computers & Graphics*, 22(6):687–695, December 1998.

139. Q. Sun and W. Hurst. Video browsing on handheld devices–interface designs for the next generation of mobile video players. *IEEE Multimedia*, 15(3):76–83, September 2008.

140. M. D. Swanson, M. Kobayashi, and A. H. Tewfik. Multimedia data-embedding and watermarking technologies. *Proceedings of the IEEE*, 86(6):1064–1087, June 1998.

141. T. Syeda-Mahmood and D. Ponceleon. Learning video browsing behaviour and its application in the generation of video previews. *ACM Multimedia*, pages 119–128, October 2001.

142. K. Tanaka, Y. Nakamura, and K. Matsui. Embedding secret information into a dithered multi-level image. *Proceedings of the IEEE MILCON international conference*, pages 216–220, 1990.

143. A. Z. Tirkel, G. A. Rankin, R. G. van Schyndel, W. J. Ho, N. R. A. Mee, and C. F. Osborne. Electronic watermark. *Digital Image Computing, Technology and Applications*, pages 666–672, December 1993.

144. T. K. Tsui, X.-P. Zhang, and D. Androutsos. Color image watermarking using multidimensional fourier transforms. *IEEE Transactions on Information Forensics and Security*, 3(1):16–28, March 2008.

145. P. Turaga, R. Chellappa, V. S. Subrahmanian, and O. Udrea. Machine recognition of human activities: A survey. *IEEE Transactions on Circuits and Systems for Video Technology*, 18(11):1473–1488, November 2008.

146. N. Vasconcelos and A. Lippman. Statistical models of video structure for content analysis and characterization. *IEEE Transactions on Image Processing*, 9(1):3–19, January 2000.

147. V. V. Veeravalli. Decentralized quickest change detection. *IEEE Transactions on Information Theory*, 47(4):1657–1665, 2001.

148. G. Voyatzis and I. Pitas. Protecting digital image copyrights: A framework. *IEEE Computer graphics applications*, 19:18–24, 1999.

149. G. Voyatzis and I. Pitas. The use of watermarks in the protection of digital multimedia products. *Proceedings of the IEEE*, 87:1197–1207, 1999.

150. H. D. Wactlar, T. Kanade, M. A. Smith, and S. M. Stevens. Intelligent access to digital video: Informedia project. *IEEE Computer Magazine*, 29(5):46–52, May 1996.

151. F. Wang, C.-W. Ngo, and T.-C. Pong. Gesture tracking and recognition for lecture video editing. *Proceedings of the 17th International Conference on Pattern Recognition*, 3:934–937, August 2004.

152. S. Wang, X. Wang, and H. Chen. A stereo video segmentation algorithm combining disparity map and frame difference. *Third International Conference on Intelligent System and Knowledge Engineering*, 1:1121–1124, November 2008.

153. Z. Wang, A. C. Bovik, and B. L. Evans. Blind measurement of blocking artifacts in images. *Proceedings of IEEE International Conference on Image Processing*, 3:981–984, 2000.

154. Z. Wang, H. R. Sheikh, and A. C. Bovik. No-reference perceptual quality assessment of jpeg compressed images. *Proceedings of the IEEE International Conference on Image Processing*, pages 477–480, September 2002.

155. Y. Wexler, E. Shechtman, and M. Irani. Space-time completion of video. *IEEE Transactions on Pattern Analysis and Machine Intelligence*, 29(3):463–476, March 2007.

156. A. Wigley, D. Mothand, and P. Foot. *Microsoft Mobile Development Handbook*. Microsoft Press, 2007.

157. A. Winslow, Q. Tung, Q. Fan, J. Torkkola, R. Swaminathan, K. Barnard, A. Amir, A. Efrat, and C. Gniady. Studying on the move - enriched presentation video for mobile devices. *IEEE INFOCOM Workshops*, pages 1–6, April 2009.

158. L. Wolf, M. Guttmann, and D. Cohen-Or. Non-homogeneous content-driven video-retargeting. *Proceedings of the IEEE 11th International Conference on Computer Vision*, pages 1–6, October 2007.

159. S. Wu, J. Huang, D. Huang, and Y. Q. Shi. Efficiently self-synchronized audio watermarking for assured audio data transmission. *IEEE Transactions on Broadcasting*, 51(1):69–76, March 2005.

160. S. Xiang and J. Huang. Histogram-based audio watermarking against time-scale modification and cropping attacks. *IEEE Transactions on Multimedia*, 9(7):1357–1372, November 2007.

161. J. Xin, C.-W. Lin, and M.-T. Sun. Digital video transcoding. *Proceedings of the IEEE*, 93(1):84–97, January 2005.

162. L. Xu, A. Krzyzak, and C. Y. Suen. Methods of combining multiple classifiers and their applications to handwriting recognition. *IEEE Transactions on Systems, Man and Cybernetics*, 22(3):418–435, June 1992.

163. C. H. Yeh and C. J. Kuo. Digital watermarking through quasi m-arrays. *IEEE Workshop on Signal Processing Systems*, pages 456–461, 1999.

164. I. K. Yeo and H. J. Kim. Modified patchwork algorithm: A novel audio watermarking scheme. *Proceedings of the International Conference on Information Technology: Coding Computing*, page 237, 2001.

165. M. M. Yeung and B.-L. Yeo. Video visualization for compact presentation and fast browsing of pictorial content. *IEEE Transactions on Circuits and Systems for Video Technology*, 7(5):771–785, October 1997.

166. P. Yin, M. Wu, and B. Liu. Video transcoding by reducing spatial resolution. *Proceedings of the International Conference on Image Processing*, pages 972–975, 2000.

167. J. Yuan, H. Wang, L. Xiao, W. Zheng, J. Li, F. Lin, and B. Zhang. A formal study of shot boundary detection. *IEEE Transactions on Circuits and Systems for Video Technology*, 17(2):168–186, February 2007.

168. R. Zabih, J. Miller, and K. Mai. A feature-based algorithm for detecting and classifying scene break. *Proceedings of ACM Multimedia 95, San Francisco, CA*, pages 189–200, 1995.

169. D. Zhang, W. Qi, and H.-J. Zhang. A new shot boundary detection algorithm. *Proceedings of the Second IEEE Pacific Rim Conference on Multimedia: Advances in Multimedia Information Processing*, pages 63–70, 2001.

170. H. J. Zhang, A. Kankanhalli, and S. W. Smoliar. Automatic partitioning of full-motion video. *Readings in multimedia computing and networking*, pages 321–338, February 2002.

171. H. J. Zhang, C. Y. Low, S. W. Smoliar, and J. H. Wu. Video parsing, retrieval and browsing: an integrated and content-based solution. *ACM Multimedia*, 1996.

172. W. Zhang, J. Lin, X. Chen, Q. Huang, and Y. Liu. Video shot detection using hidden markov models with complementary features. *Proceedings of the First International Conference on Innovative Computing, Information and Control (ICICIC)*, 2006.

173. J. Zhou and X.-P. Zhang. Video shot boundary detection using independent component analysis. *Proceedings of the IEEE International Conference on Acoustics, Speech, and Signal Processing (ICASSP)*, 2(12):541–544, March 2005.

174. X. Zhou, X. Zhou, L. Chen, A. Bouguettaya, N. Xiao, and J. A. Taylor. An efficient near-duplicate video shot detection method using shot-based interest points. *IEEE Transactions on Multimedia*, 11(5):879–891, August 2009.

175. Y. Zhuang, Y. Rui, T. S. Huang, and S. Mehrotra. Adaptive key frame extraction using unsupervised clustering. *Proceedings of IEEE International Conference on Image Processing*, pages 866–870, 1998.

Index